Advanced General Pathology of Animals

CHANDRA DEO NARAIN SINGH
B.V.Sc & A.H., M.Sc (Vet.), Gold Medalist, Ph. D.,F.R.V.C.S (Sweden)
Former University Professor-cum-Chairman of Pathology,
Bihar Veterinary College
Patna

Published by

Khushnuma Complex Basement
7, Meerabai Marg (Behind Jawahar Bhawan),
Lucknow 226 001 U.P. (INDIA)
Tel. : 91-522-2209542, 2209543, 2209544, 2209545
Fax : 0522-4045308
E-Mail : ibdco@airtelmail.in

First Edition 2010

Price: Rs. Rs 325.00
ISBN 978-[illegible]0

Composed & Designed at :

Panacea Computers
3rd Floor, Agarwal Sabha Bhawan, Subhash Mohal
Sadar Cantt. Lucknow-226 002
Phone : 0522-2483312, 9335927082, 9452295008
E-mail : prasgupt@rediffmail.com

Printed at:

Salasar Imaging Systems
C-7/5, Lawrence Road Industrial Area
Delhi - 110 035
Tel. : 011-27185653, 9810064311

Dedicated to

Lord Pawan Sut Hanuman Ji
and
to the memory of Late Veena singh,
wife of the author

संदेश

'Advanced General Pathology of Animals' के लेखक भूतपूर्व विश्वविद्यालय प्राध्यापक, डा. चन्द्रदेव नारायण सिंह, व्याधि विभाग बिहार भेटनरी कॉलेज पटना के इस पुस्तक लेखन प्रयास को जानकर अत्यन्त खुशी हुई। सामान्य व्याधि विज्ञान बीमारियों के शिक्षण, शोध एवं चिकित्सा क्षेत्र में मूलभुत संरचना होती है । मुझे पूर्ण विश्वास है कि इस पुस्तक से व्याधि विज्ञान के छात्र, शिक्षक एवं शोधकर्त्ता लाभान्वित होगें । डा. सिंह का यह कार्य प्रशंसनीय हैं । राजेन्द्र कृषि विश्वविद्यालय के कुलपति डा. एन. एल. मौर्य पुस्तक लेखन क्षेत्र में उत्साह वर्णन के लिए धन्यवाद के पात्र हैं ।

दिनांक– 18-01-2008

स्थान– पटना, बिहार

नरेन्द्र सिंह

(कृषि मंत्री)

बिहार सरकार

पटना

FOREWORD

The book "Advanced General Pathology of Animals" written by Dr. C.D.N. Singh, former University Professor cum Chairman of the Department of Veterinary Pathology deals with General Pathology and Oncology which can be profitably read by both undergraduate and postgraduate students in Veterinary Pathology. It is a matter of immense satisfaction to me that the text has been prepared in simple and lucid language which will be absorbed by both teachers and students in Veterinary Colleges.

I congratulate Dr. C.D.N. Singh for his endeavor in writing this book and wish him all the best.

Place: B.V.C. Patna
Dated: 15-02-2008

(N.L.Maurya)
Vice Chancellor
Rajendra Agricultural University
Pusa, Smastipur

PREFACE

This is a completely revised edition of the book published under the title of 'Modern approach to Veterinary Pathology-Part 1' in 1996. The subject matter of General Pathology as a whole has been synthesized in such a way to aid its teaching as per syllabi prevalent in veterinary colleges of Agricultural universities. The text of this endeavor makes pathology easy instead of making the easy difficult for the students and the teachers in Veterinary Colleges. The aim of this presentation is to acquaint the readers or users in pathology with the fundamentals of Veterinary Pathology in view of aetiology, macroscopic- and microscopic pathology of disease processes in the body of an organism. The relevance and usefulness of gross- and microscopic pathology will hold the ground for several coming decades against the advances in histochemistry, histophysics, electron microscopy (EM) and molecular pathology etc. Pathology is an observational science which is quite inseparable from gross- and microscopic pathology of diseases in organisms. Pathophysiology and pathanatomy are very useful to arrive at a proper diagnosis of diseases in the organisms without undue wastage of time and energy on the part of the students. In addition, the terms used in this book are explained in an easily comprehensible manner.

I am especially indebted to Professor Sven Rubarth, Royal Veterinary College, Stockholm, Sweden and Professor P.B. Kuppuswamy, Bihar Veterinary College, Patna for valuable encouragement and inspiration in preparation of the text. I am also thankful to Professor B. Singh, Department of Pathology, P.A.U. Ludhiana, Punjab, Dr. K.K. Singh, Head Department of Pathology, R.V.College Ranchi, Dr. Binod Kumar Singh, Head Department of nephrology, Patna Medical, College, Patna, Dr. Mohan Prasad, Head, Bihar College Physiotherapy and Occupational therapy, PMCH campus Viklang Bhawan, Kankarbagh, Patna and Dr. M.S. Kwatra, Additional Director Research PAU, Ludhiana,

Punjab, and Dr. U.K. Gupta and Dr. B.N. Prasad, Medical Officer, Health services Government of Bihar, Patna, for valuable help and incentive in completing this book. Last but by no means least, I acknowledge my gratitude to several standard works on this subject upon which I have drawn freely. I lay no claim to any originality in writing this book. My special thanks are also due to Professor Shankar Prakash, Professor, Department of Microbiology, and Patna Medical College Patna for help in photomicrography of histopathological specimens. I also express my thanks to **Neha,** Ashok Nagar Road No-1, Kankarbagh Patna-20 Shalni Raj, Patel Nagar, Patna and Somya Raj, Sector K/3, Kankarbagh Patna, Apurva Singh, Nandan Tower Colony More, Patna for preparing the type- script of this book. I am very much thankful to my publisher.

Any suggestion for improvement in the text will be appreciated by the author. My efforts will be deemed to be successful if the work is proved useful for readers and students in the field of animal pathology

D-33 P.C. Colony, Kankarbagh **C.D.N. Singh**
Patna-20 (Bihar)

Contents

Chapter 1

Introduction

Pathology (Gr. = disease + discourse) is a branch of medicine which deals with the study (logos) of changes (alterations), their causes, development, site, appearance and spread in the body of animals because of an exposure to injurious factors or deprivations. The study of disease (suffering-pathos) includes the molecular, cellular and organismal response of the individuals when exposed to injurious agents. In other words, pathology is the science of diseases incorporating the study of aetiological factors, structural and functional changes in the cells and tissues of the organs of the living body under noxious influences .

The five main aspects of a disease process (i.e. disease indicating deviations from the normal state) are the following:

(i) Causes of diseases (aetiology-science of causes)

(ii) Manners (mechanisms) of the development of the disease process in the body.

(iii) Structural (morphological) alterations in the cells and organs of a diseased animal.

(iv) Functional effects (consequences) of the morphologic changes induced in the cells and tissues.

(v) Other changes (e.g. biochemical and molecular etc.) are also noticed in the cells and tissues of diseased organs.

Animate factors (bacteria, viruses, protozoa and parasites, inanimate factors (compounds like acids, alkalies, phosphorus, arsenic etc.), mechanical or physical factors, deprivations or imbalances of proteins, vitamins, minerals, water and oxygen etc. in feeds, and genetic abnormalities are some of the important causes for producing anatomical, physiological and biochemical changes which are some changes underlying the disease processes in the animals. In short, morphological changes (changes in anatomy), biochemical changes (changes in biochemistry) and molecular changes (changes in bio-molecules) are some important factors underlying diseases processes in the body of diseased animals. Pathoanatomy and pathophysiology are the two main bases of pathology.

Pathogenesis (Gr. = disease + origin) means the evolution of a disease (i.e. the sequence of events in the cells and tissues exposed to the intrinsic and extrinsic factors till the expression of the disease as a result of the noxious influences in animals). It refers to the manner in which a disease process develops in an individual.

General pathology is concerned with fundamental processes (reactions) which are noticed in more than one organ and can often occur from more than one cause (i.e. an abnormal stimulus). In short, it discusses the basic reactions of cells and tissues of the living body to abnormal stimuli that underline several diseases. Cellular swelling, fatty changes and necrosis etc. are examples of alterations of general pathology. The term physiologic pathology deals with changes in physiology consequent upon a disease process. Changes in chemistry of an individual consequent to a disease constitute the subject matter of pathologic chemistry. Clinical pathology is that branch of pathology which is practiced in hospitals. It includes examination of blood, urine, faeces, scrapings, biopsy material and exudate etc. in the hospital. The study of a willfully produced disease in an animal is called experimental pathology. Comparative pathology is that branch of pathology which deals with comparison of diseases of lower animals with those in man. Diseases (Fr. = negative +ease) is a pathologic process marked by discord of the living body of animals with environment. In other words, it means the lack of ability of the living body to adapt itself to the irritation of existence. Diseased cells and tissues of living body of animals reveal reversible and irreversible changes. Health (As= whole) is the converse of disease which indicates a condition of body in complete harmony with its surroundings. Lesions denote the macroscopic and microscopic changes in the organs of living animals when exposed to injurious agents (i.e. animate and inanimate factors etc.). Inclusions in neurons, abscesses and ulcers are examples of lesions. Cellular swelling (parenchymatous degeneration) is considered to be a biochemical lesion in cells and tissues exposed to poisonous substances. Life means the state of labile equilibrium between divergent, simultaneous, reversible and potentially perpetual reactions in the body (Rubarth, 1964)

Molecular pathology deals with the study of biochemical and biophysical cellular mechanisms as the basic factors in diseases. Discovery of genes and molecules have greatly enriched the information on the pathogenesis of diseases. One notices a link between abnormal genes and expression of certain diseases in living bodies. Advances in the role of molecules in the pathogenesis of diseases have revolutionized the science of pathology. Nosology (i.e. the study of diseases) deals with the classification of diseases. Disease processes in particular organs of certain systems are studied under special pathology.

Definitions of some common terms

1. **Congential** (Gr. = born together) anomaly concerns an abnormality born with an individual.
2. **Hereditary** disease is a disease derived from ancestry and can be traced from generation to generation in the living body of animals.
3. **Agenesis** (Gr. = privation + generation) or Aplasia (Gr. = privation +formation) indicates incomplete or defective development of an organ or tissue in body.
4. **Hypoplasia** (Gr. = under + formation) refers to an under-development of an organ marked by failure of cells to reach their adult size.
5. **Stenosis** (Gr. = narrow) means a narrowing of a canal or opening. Atresia (Gr. = negative+boring) means an imperfection (absence or closure) of a normal opening.
6. **Ectopia** (Gr. displaced) denotes a displacement or malposition of an organ.
7. **Heterotopia** (Gr. = other +place) refers to the presence of tissues in an abnormal location. Intravascular clots in

thrombosis and islands of gastric mucosa in the oesophageal lining are examples of heterotopia.

8. **Fissure** (L. to split) means a cleft, split or groove in the tissues of the living body.
9. **Sinus** (L=bent surface or curved surface) means a blindly ending tract that extends from a mucosa of a hollow organ or from the surface of the body or an organ into adjacent tissues of living bodies.
10. **Fistula** (L = music pipe) refers to a tract connecting a hollow organ with another cavity or the exterior of the body in animals.
11. **Heterochronia** refers to a phenomenon of change in a process governed by time. Precocious lactation in a new born animal and premature (an early) aging are noticed in the cases of heterochronia.
12. **Heterometria** means an augmentation or diminution of a normal process of life.
13. **Dysplasia** refers to a developmental abnormality resulting in a change in shape of an organ. It literally means the disordered growth. The important features are lack of uniformity of the individual cells and loss in their architectural orientation, pleumorphism, hyperchromatic nuclei and more abundance of mitotic figures.
14. **Dyscrasia** means a change in the chemical nature of the processes in the body. Heterology deals with deviations from normal and affects size, time and site.
15. **Dysontology** discusses the developmental abnormalities in the living body of animals. It is exemplified by cryptorchidism. (i.e failure of the testicles to descend in the usual position).
16. **Coma** It is a clinical term (i.e. a symptom) which is marked by insensitivity which involves the brain and ends in the death of an individual.
17. **Syncope** It refers to a sudden stoppage of heart which may be fatal in an individual
18. **Granuloma** It denotes a focal area of granulomatous inflammation consisting of a microscopic aggregation of

macrophages transformed into epi[illegible]um like cells (epithelioid cells) surrounded by a c[illegible] of monocytes, lymphocytes and occasionally plasma c[illegible]

19. Pus

It refers to a purulent inflammatory exudate rich in leucocytes (mostly neutrophiles) and cellular debris. The dead neutrophiles impart a creamy yellow color to pus. The chief cellular components are neutrophiles showing degenerative and necrotic changes.

20. Ischemia (Local anaemia)

It is marked by a local diminution in the blood supply_to a part because of a decrease or cessation of inflow of the arterial blood. It is a condition marked by a diminished filling of arteries or arterial branches in the affected part of organs. Events like cellular swelling and fatty changes are noticed in the cases of reversible injuries. But necrosy is the most marked change in an irreversible injury. Cell injury in ischemia arises as a of result of reduced oxygen supply to tissues. But injurious stimuli like radiation causes cell damage by toxic activated oxygen species (O_2^-, $OH^\cdot$ and H_2O_2). Hypoxic endothelial and parenchymal cells are responsible for production of cytokines and increased expression of adhesion molecules which are associated with ischemia in tissues.

21. Ulcer

It means a local defect or excavation on surface of an organ which arises from sloughing of an inflammatory necrotic tissue. Ulcers are noticed in the mucosae of the mouth, stomach, urinary tract, vagina and subcutaneous tissue (skin) of the living bodies, Peptic ulcers develop towards the pyloric end of the stomach or gastric mucosa. In short, ulcer is a break or breach of the surface epithelium of an organ.

22. Epithelioid cells (also called Modified macrophages)

These are activated macrophages with modified epithelium (epithelioid) like appearance. In H and E stained sections of the tissues, these cel ave pale granular cytoplasm with indistinct boundaries. Ep lioid cells often tend to merge into each other. The nuclei are val, elongated and less dense when compared with those o ymphocytes.

23. Auto sis

It refers t digestion of the cells by catalytic enzymes present within ells and tissue of dead bodies.

24. Heterolysis

It means the enzymic digestion of cells by the catalytic enzymes of the infiltrating or immigrant leucocytes in the degenerated tissues.

25. Agony (i.e. a painful state of the living body)

The features of agony in diseased animals are as follows:

(i) Disordered activity of the central nervous system and disturbances in all the vital activities of bodies.

(ii) Abnormal respiratory movement, weakened heart, relaxation of sphincters, drop in body temperature and loss of consciousness.

(iii) Occurrence of agony before the clinical death of an individual. And as such, an agonal period is a time between a lethal hit and clinical death of diseased animal. With the onset of irreversible changes in the tissues or cells, a true death sets in the body.

26. Cyst

Cyst is a structure having a cavity lined by an epithelium with a surrounding wall of fibrous tissue. Fluid or semifluid material is found in the cyst. Different terms indicating the varieties of cysts included dermoid cyst, dentigerous cyst, parasitic cyst and cystic forms arising from an interference with the usual escape of secretion from the saliary glands.

Reversible and irreversible injuries

Injurious factors (e.g. bacteria, viruses, chemicals, heat, cold, light and radiation) cause damage to cells or tissues in the living body of animals. Both local and general reactions arise from the damage in tissues of the body. The damage produced in the tissues of living body is called injury. Injurious agents produce abnormal changes (called lesions-deviations from normal) in different parts of the body. Abnormal changes in tissues or organs produce deviations from the normality in the cells or tissues in the injured patients. Tissues damaged (i.e. muscles) by trauma reveal morphologic changes and potassium, creatine, phosphate and amino nitrogen are released into the blood stream from the damaged tissues. Ischemia can be produced by constricting the limbs. Animals can be injured by scalding which gives rise to structural and biochemical abnormalities.

When adaptive response to an abnormal stimulus is not adequate, one notices a sequence of events (called injury) in the irritated tissues. The sequence of events (i.e. morphologic changes) may be reversible and irreversible. Reversible sequence of events in the damaged tissues is called reversible injury where as sequence of events up to the point of no return gives rise to irreversible injury in the tissues ending in the ultimate death of cells.

The main features of reversible and irreversible injuries are given in **Table-1.** Irreversible changes characterize coagulation necrosis (circumscribed necrotic areas) because of hypoxic death following lethal ischemia are seen in embolism and thrombosis in the blood vessels, in organs like kidneys and heart etc. Cardiac noxious influences in the ischemic zone are first seen at molecular levels. Cellular injury is followed by appearance of detectable ultrastructural and histochemical changes within a few minutes to hours. But gross and microscopic changes are noticed in the areas of coagulation necrosis alter a long period of time. The main causes of cell injury include hypoxia and ischemia in the tissues. Hypoxia arises from insufficient supply of oxygen to the tissues and its inadequate utilization by them. Embolism and thrombosis cause partial or complete occlusion of the blood vessels and the condition of ischemia in tissues is produced. Injuries in tissues in body also arise from reduced venous drainage from a part, chemicals, physical agents, bacteria, viruses, immunologic reactions (e.g. anaphylaxis) and genetic derangements or nutritional imbalances of fat, protein, carbohydrate, minerals, and vitamins in feeds.

Table: 1. Some features of reversible and irreversible injuries.

Reversible injury	Irreversible injury
1. Reversion to the state of normality. Recovery of cells on availability of suitable conditions e.g. oxygen supply to tissues.	1. No reversion to normality as a result of irreparable damage to cell membranes, genome and depletion of ATP in cells.
2. Loss of oxidative phosphorylation, (aerobic metabolism) and decreased production of ATP. Hypoxic injury is marked by such change.	2. Persistence ischemia followed by irreversible injury marked by swelling of mitochondria, damage to plasma membranes, swelling of lysosomes and large amorphous densities in the mitochondria
3. Reduction of Na+K+-ATpase i.e. plasma membrane energy dependent sodium pump severely affected	3.Increased influx of calcium in cytosol of the affected cells, and decrease in membrane phospholipids
4. Decreased ATP concentration and enhanced AT Pase activity	4. Escape of proteins, enzymes and ribonucleic acids through the permeable cell membranes.
5. Influx of sodium, water, chloride in the cells, enhanced ATpase activity and decreased concentration of potassium in the cells.	5. Leakage of cellular enzyme from the injured lysosomal membrane in cytoplasm and activation of the released acid hydrolases. Loss of basophilia due to decrease in RNP (ribonucleoprotein)
6. Cell swelling and dilation of the endoplasmic reticulum.	6. Clumping of nuclear chromatin because of decreased cell pH value and large masses of myelin figures consisting of phospholipids.
7. Increased anaerobic glycolysis to generate ATPase metabolism of glucose derived from glycogen.	7. Formation of calcium soaps with fatty acids and calcification of fatty acid residues.
8. Decrease in the intracellular pH due to production of lactic acid and inorganic phosphates.	8. Damaged cell membranes and cell constituents damaged by reactive oxygen species.
9. Decreased protein synthesis and detachment of ribosomes from granular endoplasmic reticulum.	9. Discernible tissue structures, coagulated acidophilic cytoplasm, anucleated cells, fragmentation of nuclei, irreversible nuclear changes like pyknosis, hyperchromatosis karyorrhexis and karyolysis and phagocytosis of cellular debris are some important light microscopical changes
10. Loss of microvilli and formation of blebs at the cell surface, myelin figures derived from plasma and membranes of cell organelles are observed as important ultrastructural features woresening cases of hypoxia.	10. Lysis of endoplasmic reticulum and development of defects in the cell membranes . Ultrastructural changes include shrunken nuclei, more electron dense chromatin than normal, hydropic changes and myelin figures in the cytoplasm, chromatin fragmented into electron dense masses and distruption of nuclear and cellular membranes.

Death

Thanatology refers to a discourse on the death in all its forms in the living body.

The three types of death are :
(i) Clinical death
(ii) Somatic death (i.e. systemic)
(iii) Molecular (cellular) death

Clinical death

It denotes the deepest state of the brain depression (CNS) in the living animals. The changes in the tissues are still reversible but the changes pass into irreversible changes after 5 to 6 minutes.

Somatic death

A complete and irreversible stoppage of the respiration, circulation and brain function which leads to death is known as somatic death. Man and animals are alive till the continuation of oxygenated blood supply to the brain stem continues without any break. Transplantation of cornea, skin and bone is carried out after few hours following death. Animals are declared dead when there is cessation of heart beat, respiration and brain activities. An irreversible loss of cerebral functions is called brain death of an organism.

Molecular death

This term refers to death of all cells and tissues individually. Neurons die quickly in about five minutes (say, as found in humans). But the muscles survive up to 1-2 hour. Anoxia (lack of oxygen in the blood stream) stops metabolic activity in cells and tissues and vital functions of organs like brain and heart etc. Anoxia is noticed in the cases of drowning, hanging and strangulation etc. Necrosis reveals changes like cellular swelling, cell rupture, coagulation of cellular proteins and break down of cell organelles in certain kinds of injuries e.g. irreversible ischemic, chemical and radiation injuries. Information in relation to reversible and irreversible effects of deficient blood supply to tissues is given below. (Table-2)

Table- 2. Main features of the arrest of blood supply or deficient blood supply to tissues

Tissues	Duration of the arrest	Pathologic effects
1. Cardiac muscle	10-15 minutes	Reversible degenerative changes like cellular swelling, fatty changes and necrosis etc.
2. Cardiac muscle	1 hour	Irreversibble degenerative changes (i.e., cellular death)
3. Brain cells (neurons)	3 to 7 minutes	Do
4. Skin cells (epidermal cells)	Up to 24 hrs	No effects following arrest of blood supply for a few minutes
5. Bones	Up to 24 hrs	Bones removed after 40 hours from dead bodies successfully transplanted.
6. Hairs and nails	Up to 24 hrs	Growth of hairs and nails (say, in humans) is noticed.

Thus, more endurance to oxygen arrest is shown by certain tissues like skeletal muscle, connective tissue and skin depending on their susceptibilities. Apoptosis and necrosis of tissues are two kinds of the cell death in the living body.

POSTMORTEM EXAMINATION

Postmortem examination (L. =after death) refers to both the external and internal examinations of the body after death. The term postmortem examination is synonymous with terms like autopsy (Gr. =self + view), necropsy (Gr.= dead+view) or necroscopy (Gr. =corpse +to view), Its purpose is to ascertain the cause of death and study the nature, extent and evolution of the disease affecting an organism. Studies of different kinds like gross anatomic examination (first with the naked-eye), microscopic anatomical examination (with the aid of a microscope), bacteriological, virological and chemical examinations are made of the different organs and materials collected from the dead body in order to arrive at a definite diagnosis (aetiological or biological etc.) The autopsy is performed by adopting a uniform and careful procedure. The organ systems are removed from the dead body with caution and maintenance of the anatomical continuity of organs within the systems. The organs from the body must be removed in a correct way.

Anamnesis (history) of the dead animals must be thoroughly studied and if necessary, more details of the dead animals should be obtained from the owner of the animals. Information like place and date and time of death, the species, breed, sex, age, distinguishing marks, the presence or absence of rigor mortis, abdominal tympany, rectal protrusion and an odour of putrefaction etc. are very essential for making diagnosis of the diseases affecting the animals. The postmortem examination should be made soon after death in order to make a quick diagnosis. A long dead animal is not suitable for postmortem examination due to growth of putrefactive organisms. Highly decomposed bodies of animals are useless for the bacteriological examination.

HISTORY OF PATHOLOGY

Pathology (a specialty of a branch of medical science) refers to the study of the causes and structural and functional defects produced by these factors in the body of an organism. Its evolution took place gradually and is based on the observations or contributions of several workers of ancient civilizations up to the present time. The peoples like Egyptian, Jews, the Assyrians, Chinese and Indians of the olden times were aware of several diseases as evident from the survey of ancient writings or paintings. It is quite apparent from an Egyptian papyrus (about 2160-1778 B. C.), Edwin Smith papyrus (1600 B.C.) and Ebers papyrus (about 1500 B.C) that the diseases like coryza, dysentery, tumours, cystic conditions and abscesses etc. were known to these ancient peoples. The pathological science saw the days of remarkable progress from the period of Renaissance (14th to 16th centuries). There was much progress in physics, chemistry and biology. Biologists used it and the knowledge in pathological science was greatly enriched by the naked-eye and microscopical observations of the postmortem lesions. Histochemistry and histophysics followed the disease processes into the cells or tissues of the organs of sick individuals. Now-a-days, electron microscopy has further helped pathological studies in the cellular organelles e.g. nuclei and mitochondria etc. of the affected organs of diseased animals.

People became aware of the parasites or ulcerating masses by 1550. In 5th century B.C., Hippocrates (460-370 B.C.) considered the diseases due to disturbances in the body fluids (humors). An improper mixing of these fluids (e.g. blood from heart, phlegm from brain, yellow bile from liver and black bile from spleen) caused the diseases and, thus, the theory of humoral pathology was established. A text book on special pathology was written by Cornelius Celsus (30 B.C.-38 A.D.) and it was he who described the cardinal signs like rubor (redness), tumor(swelling), color (heat) and dolor (pain) characterizing the inflammatory processes. Galen (130-200 A.D.) noted the fifth cardinal sign of inflammation known as loss of function (functio laesa). There is a recorded history of veterinary medicine during the reign of Hamurabi (2100 B.C.). A little progress was made in pathology up to 14th to 16th century (i.e. the period of Renaissance). Cohnheim (1882) gave an account of events seen during inflammation in web and tongue of the frog following locally induced injury.

Antonio Benivieni (1440-1502) was the first pathologist to be credited with postmortem examination of human corpses and has been accepted as father of pathological anatomy.

The discovery of magnifying glasses or microscope greatly helped the study of the changes in the cells or tissues of diseased individuals. Antony van Leeuwenhoek (1632-1723) showed the importance of microscope for the study of tissues or minute objects. Capillary circulation and blood cells in the body were discovered by Malpighi (1628-1694). Giovanni Battista Morgagni (1682-1771) correlated postmortem findings with the symptoms noticed in diseased individuals. Records of the history of patients and the diseases affecting them were made by him together with events in connection with the illness. Correlation of the symptoms in the body with the lesions produced therein is a very important step in the modern medicine. John Hunter (1728-1793) greatly advanced museum techniques in the pathological science.

Mathew Baillie (1761-1823) wrote a book on pathological anatomy arranging the matter according to the organs. Bichat (1771-1802) noted the organs consisted of tissues which were divided by him into 21 kinds by treating them with acids, alkalies, salts, heat and air etc. He, further, divided pathology into two parts, namely (i) general pathology (alterations common to the organs of each system) and (2) special pathology (diseases or alterations peculiar to each organ) Laennec (1781-1826) invented the important diagnostic tool called stethoscope which greatly enhanced the knowledge on pulmonary tuberculosis and emphysema etc.

Carl Rokitansky (1804-1878) was a very renowned autopsist in medical science who was credited with performance of 30,000 autopsies and handling of 60,000 pathological materials. Claude Bernard (1813-1878) established the foundation of experimental pathology.

Schwann (1810-1882) described the cells in the somatic structures with the use of microscope and the cellular theory, thus, come into existence. R. Virchow (1821-1905) pointed out towards the alterations or changes in the cells of diseased organs and considered the pathological processes as the sum total of the cellular changes. The foundation of the cellular pathology was so laid by this noted pathologist who also started medical journal known as Virchow's Archives. A good deal of progress in the morbid and microscopic pathology was made in the 19th century. Electron microscope facilitated further observations, not detectable by light microscopy, in the different ultra-microscopic structures of the cells. Pasteur's and Koch's several discoveries in the 2nd half of the 19th century revealed the animate or viable factors (e.g. bacteria and viruses) to be responsible for diseases in the affected individuals.

Mechnikov (1845-1916) noticed the defensive mechanisms of body by discovering the protective cells, like neutrophiles (microphages) and monocytes or hisiocytes (macrophages) in the inflamed tissues caused by pathogenic microbes.

Many veterinary schools or colleges were established in different parts of the world. The first modern veterinary school was established in Lyon (France) in 1762 and the veterinary college of London came into existence in 1791. Later, many British veterinary schools at Toronto were established by Andrew Smith. Albert Hajare and Seven Rubarth (Sweden) were renowned Scandinavian veterinary pathologists. Several Indian veterinary pathologists associated with teaching and research in pathology at institutions in places like Izatnagar, Mathura, Madras, Kerala, Bengal, Bombay and Jabalpur etc. have done very good work on animal diseases to be absorbed by the students of veterinary pathology.

Dr. C.M. Singh, Dr. B.S. Rajya and Dr. P.K. Iyer (I.V.R.I.., Izatnagar), Dr. J.L. Vegad (Jabalpur) Dr. P.B. Kuppuswami (Patna), Dr. B. Singh, (Ludhiana), Dr. M.K. Bhowmik (Calcutta), Dr. M.S. Kwatra and Dr. P.P. Gupta (Ludhiana), Dr. Nair (Kerala), Dr. S.M. Azinkya and Dr. S. D. Pandey (Bombay), Dr. S. Damodaran (Madras), Dr. G.A. Shastri (Triputi), Dr. B.C. Nayak and Dr. A.T. Rao, (Orrisa), Dr. H.V.S. Chauhan (Ranchi), and Dr. D.S. Kalra (Hissar) are some of the noted Indian Veterinary Pathologists who have made valuable contributions in elucidating many disease problems in animals in this part of the Asiatic sub-continent.

AETIOLOGY

Aetiology (from the Greek words, aitia-causes and logos-words or account). Aetiology deals with the causes of disease in an organism. The causes of diseases are divided into two types:

1. Intrinsic or predisposing causes e.g. age and sex, etc.
2. Extrinsic or direct causes

Two main kinds of extrinsic causes are :

(i) Animate factors (e. g. bacteria, viruses, metazoan parasites and protozoa etc.)

(ii) Inanimate factors (e.g. chemical factors like carbon tetrachloride, phosphorus and arsenic etc.)

Intrinsic or predisposing causes of diseases

There are several intrinsic causes of diseases. These factors themselves do not directly produce any disease but produce unusual susceptibility of an individual to a disease or even determine the type of the diseases present within an individual. Predisposition is a latent state which is affected by the extrinsic factor the production of a disease. Predisposition is inseparable from life. The risk of developing a particular disease may be associated with a certain age, sex, genus and breed etc. The lowered resistance of the organs is a condition for onset of the typhoid infection. Inadequate nutrition, poor housing and humid climate etc. are conductive for development of tuberculosis.

Pigeons which are usually unsusceptible to anthrax infection, can contact this disease if they are preliminarly subjected to starvation (a debilitating factor).

Some details of different predisposing or intrinsic factors are as under :

1. Age

Young animals are greatly susceptible to many infectious diseases. This is exemplified by high incidence of tuberculosis in children. Since animals are seldom allowed to live to senility, age predisposition is not well understood in many species of animals. However, in dogs, there is an age predisposition. Older animals frequently show tumours. Sarcomas occur in young animals but cancers are noticed in old animals. Young dogs (mostly before a year of age) suffer from canine distemper. Strangles are noticed in young horses. Anaemia in pigs due to deficiency of iron is noticed in the 3rd and 6th week of their life. Enlarged prostate due to hyperplasia is noticed in old dogs, man and seldom found in young ones.

2. Breed

Tumours of Danish mice cannot be transplanted into English or German mice. Gliomas are noticed in the brachycephalic breeds of dogs (Boxers) and degenerations of intervertebral discs occur in chondrodystrophied breeds of dogs. There is a high incidence of brain tumours in boxer dog.

3. Sex

Mammary cancer is noticed in women. Tuberculosis and brucellosis are more frequent in testicles than in the ovary. Adenoma of circumanal glands is found in male dogs. Diseases like mastitis, milk fever, metritis and acetonaemia etc. are noticed in females, Nephritis is more common in cows.

4. Organ

Certain viruses and parasites show organotroprism. Parasites have their own localization in the body. Tuberculosis is seldom found in skeletal muscles or thyroids, but it is often found in the lungs and liver. *Br. Abortus* prefers organs like testicles, uterus and foetal membranes. Rabies virus shows affinity for the nervous system.

5. Genus / Species

There are some diseases affecting certain genera. Hogcholera is found in hogs. Canine distemper primarily affects dogs. The virus of the swine fever does not affect a horse or cattle.

6. Colour

Grey and white horses commonly suffer from melanosarcoma. Animals with nonpigmented skin are very susceptible to photodynamic diseases. White bull dogs are often deaf due to lack of pigments.

7. Race and family

The susceptibility of dairy cattle is more than that of beef cattle. Certain families of animals have various lethal and abnormal characteristics. Chickens with unusual resistance to leucosis can be produced and there are certain families of chickens with great mortality rate from this disease. Certain herds of cattle have incidence of hydrocephalus (fluid accumulation in the ventricles of brain).

8. Idiosyncrasy

It refers to an unusual state of reaction to some substances like drugs or foods. When an usual recommended dose of drugs is given to an individual, he shows severe manifestations of toxicity and may even die.

9. Climatic factors

Seasonal predisposition is noticed in an individual throughout the year. Certain epizootics or epidemics reach at their peak at a particular season. In temperate regions of the world, some diseases are severely present during the spring.

Extrinsic causes of diseases

These are indispensable factors for development of diseases. In short, a direct cause is a *sine qua non* for occurrence of a disease. The main extrinsic causes of diseases are as under :

(1) Biological factors
(2) Chemical factors
(3) Radiant factors
(4) Electricity
(5) Thermal factors
(6) Mechanical and traumatic factors
(7) Nutritional factors
(8) Atmosphere

Injurious (noxious) agents damage the cells and tissues in the living body of organisms and induce a sequence of events of reversible or irreversible degenerative changes in them. This sequence of abnormal morphologic changes (pathoanatomy) is called injury (in loose sense) in the damaged and irritated tissues. Cellular swelling and fatty changes in the damaged tissues are reversible changes where as necrosis (cell death) is an irreversible change in the body when exposed to noxious influences. When the reversible degenerative changes reach point of no return in the cells, the consequence of such changes is cell death. Necrosis and apoptosis are examples of two kinds of cell deaths in the living body.

(1) BIOLOGICAL FACTORS

Diseases are caused by several biological factors (e.g. bacteria and viruses etc.). Infection means an implantation of living pathogens or microbes in an organism. Diseases arise from interaction of the individuals body and infectious agents. Virulence of the infective agent and susceptibility of the host are very important factors in production of a disease. The manner in which a disease develops in the living body is pathogenesis and the pathogenicity (i.e. capacity to produce the disease) is the main property of the causative agents of the infectious diseases i.e. they must be able to exert toxic effects. Microbes become virulent by producing aggressins, antipeptolytic enzymes and hyaluronidase which break the hyaluronic acid, a constituent of connective tissue and this leads to increase in the permeability of the tissues. Development of purulent processes (i.e. pyaemia) in different parts of the body occurs due to spread of infective agents from one part of the body to the other. Presence of the infective agent in blood is called bacteraemia and it spreads throughout the body of the organism along with fever and petechiae due to toxic products in the blood and produces a state of septicemia.

The pathogenicity of the organisms depends upon toxins i.e. exotoxin and endotoxin. Exotoxins are substances liberated by the infective agents and these are absorbed to affect the entire organism and bacterial intoxication is produced. When bacteria are destroyed or destructed, endotoxins from the bodies of the organisms are released. Powerful endotoxins are produced in typhoid fever and cholera. Presence of toxins in the body is called toxemia. Living pathogens like microbes, viruses, parasites and protozoa etc. produce different diseases in animals; Sapraemia refers to the presence of putrefaction products in the blood. Putrefaction results in the formation of toxic substances due to break-down of organic material through the activity of the microbes and anaerobic bacteria like *Clostridium* and *Proteus* spp. etc. Due to putrefaction in dead tissues, neurin and various ptomaines are produced.

Parasites refer to members of plants or animal kingdom which can exist on human and animal bodies deriving their nourishment from their hosts. Obligate parasites are solely parasites. Facultative parasites are capable to exist on the body occasionally or under particular circumstances in the parasitic form. Saprophytes are organisms existing in non-living matter i.e. putrefying material. Non-pathogenic parasites do not produce disease but they can vegetate on their hosts in an innocuous state and can be found in the skin, oral cavity, forestomach of ruminants, intestines, large intestines, respiratory tract, vagina, cisterns of the mammary glands and prepuce in the state of symbiosis. It means the relationship between the parasite and host is beneficial. Micro-organisms in the digestive tract fascilitate digestion (particularly in herbivores). When the relationship between the parasite and host is a disadvantageous one, the process of disease is produced. Micro organisms (the lowest forms of animals and plants) produce diseases and are so known as pathogens. Infectious disease is an abnormal state which is caused by micro-organisms from the plant kingdom. Infestation refers to an abnormal state in the body caused by pathogenic forms from the animal kingdom. Circulating fluids like blood and lymph are often sites of parasitism.

The effects of parasites are as under :

1. Mechanical effect in the form of an occlusion often seen in blood vessels.
2. Consumption of oxygen meant for the host.
3. Production of toxic substances.
4. Production of antibodies.

Intoxication is a pathological state produced in the body by toxins of the organisms. Bacterial intoxication can produce diseases like anthrax, tetanus and diphtheria. The endotoxins are released when the bacteria die and disintegrate. Intoxication due to endotoxins is seen in cholera, salmonellosis and typhus.

The anti-bodies against toxins are called antitoxins. The antitoxins are specific for each bacterium.

The mucoprotein produced by bacteria attracts neutronphiles. The injection of dead organisms in the body causes suppuration due to mucoprotein. Mucoprotein shows chemotaxis (chemical attraction) for neutrophiles. Haemolysin produced by bacteria causes haemolysis. The leucocidin produced by bacteria destroyes the leucocytes. Infections or diseases can be caused by two or several micro-parasites attacking the tissues simultaneously and account for the pathological processes in the body. Such an infection is called a mixed infection. Infection with one micro-organism can pave the way for the penetration of another micro-prganism which produces the condition called secondary infection. Double infections are very common and give rise to complicstions of infections. Secondary infecting organisms are cocci, streptococci etc. and the secondary invaders may be the cause of death. Infected individual can spread the infections to others by contact of such diseases. Sometimes, pathogenic micro organisms may be present in human beings or animals without any overt or clinical manifestations of the diseases and such infected individuals or organisms are called carriers. Many infective agents are ubiquitous in the environment e.g. tetanus organism existing as a facultative parasite. But obligate parasites do not exist in the external environment except the living body. The infective agents' penetrate into the body through the skin, digestive tract, respiratory organs, urogenital organs and by direct administration into the body.

Bacteriaemia leads to metastasis (growth of the bacteria at the secondary sites). In septicemic infection, the blood is dark, imperfectly coagulated in the dead animals which show petechial spots on serous and mucous membranes. Red cells may be haemolysed due to haemolysins produced by the bacteria. Infections arise from the penetration or entry of pathogenic micro-organisms through the skin or mucous membrane into the body. Damage to the body is mediated by the bacterial toxins. Many bacteria *(e.g. Anthrax bacillus)* reduce the resistance of the host by producing substances called aggressins. In viral infections, infections pave the way for infection with bacteria like streptococci. A bacterial autointoxication arises from infections of the intestinal tract. A latent period of a micro-organism is the period during which bacteria are present in the body without showing any symptom of the diseases. And an enzootic disease is a disease limited to a small area. If it involves larger areas, it is called an epizootic. A panzootic disease refers to a disease which involves several countries. In generalized infection, the infective agent spreads everywhere in the body through the blood vessels. But in localized infection, the infective agents are found at a particular site and produce lesser effects than when administered subcutaneously or into the blood. The exogenous substances may be i.e. either organic or inorganic in nature.

2. CHEMICAL FACTORS

These substances produce various kinds of effects and often cause poisoning by combining with the protoplasm of cells so as to produce structural and chemical changes in them. Intoxication and toxicosis are actually poisonings. Poisons which enter the body from the outside (or exterior), are known as exogenous substances and poisons produced inside the body are called endogenous poisons which are metabolites or products of tissue decomposition (e.g. burns and gangrene.) Autointoxication (e.g. icterus) is self poisoning due to such autogenous poisons. It may develop from abnormal processes of absorption from intestine and metabolic disorders as seen in infectious diseases and diabetes mellitus and dysfunction or excretory disorders. Uraemia results from suppressed function of the organs (e.g. kidneys) of the body. Exogenous poisoning depends upon the dose of the substance, mode of administration and resistance of the organism. The effects produced by poisons in the body may be therapeutic, toxic or lethal ones. Poisons absorbed through the gastrointestinal tract are partly or completely detoxified.

To sum up, there are certain mechanisms adopted by chemicals, drugs and toxins to induce cell injury in the body.

These are as under :

(i) Chemicals (e.g. exogenous chemicals) bind to some critical structural components or cellular organelles. For example, mercuric chloride combines with sulfhydryl groups of the cell membranes and other proteins. This leads to increased membrane permeability and inhibition of the ability to produce the ATP. Normal homeostasis and function of the cells are greatly disturbed in such reactions. Mitochondrial, cytochrome oxidase is poisoned by cyanides and oxidative phosphoyrylation is blocked.

(ii) Certain chemicals are first converted into reactive toxic metabolites which hit the target cells resulting in irreversible membrane damage and cell death (necrosis) in the living body. Carbon tetrachloride is converted into reactive toxic free radical CCL3(CCL4 +e $\rightarrow$ CCL3 + Cl^-) by P-450 mixed function oxidases. The free radicals produced cause auto-oxidation of polyenic fatty acids present within the membrane phospholipids. Lipids undergo oxidative decomposition and organic peroxides are formed following reactions with oxygen (lipid peroxidation). This kind of reaction is autocatalytic leading to formation of new radicals from the peroxide radicals themselves. A break down is noticed in the structure and function of the endoplasmic reticulum.

The effects of CCl 4- induced hepatic injuriy are given below .

(i) Fatty liver

It arrises due to membrane damage, polysome detachment and decreased apoprotein synthesis

(ii) Hepatic necrosis

It is marked by sequence of events like release of products of lipid peroxidation, damage to plasma membrane, increased permeability to sodium Na+, H_20 and Ca^{2}+, cellular swelling and rapid influx of Ca^{2}+in the cytosol. Increased concentration of intracellular calcium is toxic to cells.

Inorganic (exogenous) poisons

Acids, alkalies, salts of lead, mercury, arsenic, copper, chlorine, iodine, zinc sulphate and carbon-monoxide are examples of inorganic poisons. Corrosives act locally resulting in changes like burns, necrosis and ulcer etc. The affected tissues may be coagulated or liquified, and there will be sloughing of dead tissues leaving behind an ulcer which may heal with formation of a scar. Phosphorus and arsenic etc. produce the following kinds of effects :

1. General toxic effects of chemicals e.g. cyanide compounds or narcotics.

2. Effects on the blood

For example, potassium chlorate and carbon monoxide may be considered in this respect. Haemoglobin combines with carbon monoxide and is rendered incapable of combining with oxygen. The affinity of carbon monoxide to combine with haemoglobin is 200-300 times more than that of oxygen.

3. Effects on the liver

For example, carbon tetrachloride, phosphorus, arsenic and phlorhizin cause degenerative changes in the liver.

4 Asphyxiating effect.

For example, phosphogene and chlorine produce such effect.

5. Effects on nervous tissue

For example, toxic effects of strychnine and arsenic etc. affect nervous tissues.

Repeated administration of chemical substances (poisons) frequently result in habituation which may arise from diminished permeability of the skin and mucous membrane as seen in the case of arsenic poisoning. Habituation may result in the faster destruction of alcohol. Defensive physiological reactions may develop against poisons like ricin or arbin and the tissue may have reduced sensitivity to such poisons. There may be also faster excretion of poisons as noticed in cases of atropine. Repeated poisonings may cause allergic states.

Organic poisons

These include substances like phytotoxins, alkaloids, saponins and glucosides (poisons of vegetable kingdom) and snake venom, cantharidin etc. of animal orgin. Animal alkaloids, ptomaines and products of putrefaction are also organic poisons. Hydrocyanic acid poisoning results from eating the sorghums. Alkaloids affect the central nervous system. Hydrocynic acid inhibits the activity of the cell enzyme cytochrome oxidase resulting in disturbance of cell respiration. Saponins are glycosides that cause haemolysis of red cells. In sweet clover poisoning, commarin released from the glycosides becomes dicoumarin (an anticoagulant) which causes prolonged bleeding. Oleander contains a cardiac glycoside. Poisons like oxalic acid in halogeton, excessive amounts of nitrates and selenium in plants, molybdenum in leguminous plants, ricin, coniine in hemlocks, phytotoxins in castor bean and prophyrins in plants produce disorders (poisonings) in animals which ingest them. Body defenses against poisons include vomiting, diarrhea (due to increase in peristalsis), and excessive mucous formation by the irritated mucosae and excretion by way of the kidney and lungs etc.

3. RADIANT FACTORS

Radiation injury includes electromagnetic waves and fluxes of high speed particles of matter (electrons and protons etc.). Visible light has a wave length of 0.4-0.75 micron whereas as infrared rays and ultraviolet rays of solar system have got thermal effects and produce hyperemia of skin. Ultraviolet rays have short wave lengths and produce hyperemia of skin, pain, exudative inflammation or eczema in photosensitive subjects. These rays produce drop in blood pressure in 6-12 hours in human beings and can even change metabolism. Proliferation of cells (thickening of the epidermis) and deposition of pigment (melanin) are caused by these rays. These rays can cause death of bacteria and protozoa etc. Shock is produced by ultraviolet rays. (with wave lengths of 0.4 to 0.1 micron).

Ionising radiations include the following rays.

1. Alpha rays
2. Beta particles
3. Gamma rays.

These rays differ in physical and biological effects. X-rays arise when atoms are bombarded by high speed electrons. All these kinds of radiations cause ionization in the tissue on absorption. Inflammatory changes are produced by infrared rays with a wavelength of 0.75 to several microns. During rapid decay of atomic nuclei, there is a flux of neturons (neutral particles) and protons (positive particles). Alpha particles have low penetrating power affecting deeply located tissues on introduction of radio-active substances. Gamma rays have weak ionizing but considerable penetrating power. Gamma rays, X-rays and neutrons usually affect the entire organism. Alpha rays, beta rays and slow neutrons possess slow penetrating power and affect parts of the bodies which are exposed to them. Injury caused by radiation to the body may not be immediately apparent, but after passage of some latent period, changes or mutations in genetic material, mitochondria, Golgi apparatus, fragmentation of the protoplasmic molecules and polymerization of smaller molecules can be seen. Ionising rays damage the young growing cells in the state of mitosis. Prolonged exposure of skin to X-rays causes chronic ulcers, cancerous growth, dystrophy of cells and disturbance in the enzyme system. In ionizing radiation, there is an ionization of water molecules in the organism with formation of H, OH and H_2O_2 radicals. HO_2 and H_2O_2 radicals disturb the enzyme system. The order of sensitivity of the different cells is as follows :

1. Lymphocytes and germinal cells
2. Granulocytes and red cells
3. Epithelial and endothelial cells
4. Smooth muscle cells
5. Fibroblasts and their derivatives
6. Neurons

Endothelial cells are very sensitive to radiations causing vascular changes. Radiation injury produces blood clots, dilation of the capillaries, hyalinization of the connective tissue, fusion and homogensiation of the collagen fibres. Necrotic and ulcerative changes are produced in the body. Red cells, leucocytes and platelets are very much affected by the ionizing effects.

Thrombocytopenia is found in the irradiated individuals. Strontium, radium, uranium and zinc isotopes are strong radioactive substances and cancers or sarcomas are produced by them on introduction into the body of an organism.

4. FREE RADICALS

A free radical is a chemical species having a single unpaired electron in its outer orbit. The instability and reactivity of the free radical anions arise because of an unpaired electron in their outer orbits. Free radical processes in the living body are chain reactions (i.e. autocatalytic reactions) marked by generation of other free radicals. The chemicals or molecules with which the free radical react are themselves converted into free radicals to propagate the chain reactions of the cell injury i.e. damage to cells. The different terms like reactive intermediates, activated oxygen species and toxic intermediates refer to free radicals. The radicals react with inorganic and organic chemicals e.g. lipids, proteins and carbohydrates etc. in membranes and nucleic acid in the cells. All such chemicals cause lipid per oxidation of cells and organelle membranes, inactivate cellular enzymes and produce damage to DNA. This result is an irreversible damage to cells. As a consequence of such cellular damage, there is an influx of water, calcium and inorganic chemicals in the cytosol. The affected cells show swelling, reach the point of no return i.e. the changes of necrosis. Such necrotic changes are also noticed in hypoxic cell injuries.

The main free radicals are:

(i) Super oxide anion radical (O^{2}-)

(ii) Hydrogen peroxide (H_2O_2)

(iii) Hydroxyl ions ($OH^{\cdot}$)

Exposure to radiation e.g. ultraviolet light and X-rays, causes intracellular oxidative reactions in organs. Enzymic degradation of exogenous chemicals or drugs leads to generation of free radicals in the cells. Some details of these processes are as under:

(i) Ionisation radiation (say, ultraviolet light and X-rays) hydrolyses water into hydroxyl (OH^0) and hydrogen ($H^{\cdot}$) free radicals in the cells.

(ii) The reduction-oxidation reactions are normal metabolic processes in the cells. Normal respiratory activities are marked by reduction of molecular oxygen by addition of 4 electrous to generate water. In doing so, small amount, of toxic intermediates like superoxide (O^{2-}) hydrogen peroxide (H_20_2) and hydroxyl ions (OH^0)are also produced. There is a rapid burst of superoxide production (0_2^-) by activated polymorphonuclear leucocytes in inflammatory reactions. Even some intracellular oxidases (e.g. xanthine oxidase) produce superoxide anion radical (0^{2-}) as a result of such activities. These oxidatives generate reactive oxygen species (O^{2-}) from 0^2 in the endoplasmic reticulum (ER), mitochondrial plasma membranes, peroxisomes and cytosol.

(iii) The super oxide (O^-_2) is converted into $H_2\,0_2$ (hydrogen peroxide) by dismutation due to an enzyme superoxide dismutase. In Fenton reaction, hydrogen peroxide in the presence of ferrous form (Fe2+) of iron is transformed into OH. ($H_2\,0_2$ ® OH·+OH$^-$ ® H^2O). Three toxic intermediates namely, (1) super oxide (O_2^-) produced by autooxidative reactions in mitochrondia and oxidase reactions in the cellular cytoplasm, (2) hydrogen peroxide (H_2O_2) by dismutation of O_2^- by superoxide dismutase and (3) hydroxyl ions (OH·) released by ionization of water by radiation damage the cellular organic compounds at molecular or organellar level. The damaged molecules by these toxic products include (1) macromolecules (nucleic acids and proteins), (2) supramolecules (ribosomes and lipoproteins) and (3) cell organelles (mitochondria, lysosomes, vacuoles, Golgi bodies and nuclei – the largest cell organelle). The scavanging (neutralising) effects against toxic intermediates are produced by enzymes life superoxide dismutase, catalase and glutathione peroxidase in the injured or inflamed tissues or at the sites of elaboration of such toxic materials.

(iv) Nitric oxide (a chemical mediator) is generated by macrophases, neurons and endothelial cells in the body. It acts like a free radical which is converted into reactive peroxynitrite anion ($ONOO^-$), NO_2 and NO_3^-

The main pathologic effects of free radicals are as follows:

(i) Lipid peroxidation of membranes is caused by free radicals

Like hydroxyl ions (OH^0) lipids in cell and organelle membranes undergo peroxidation. The peroxides are produced by lipid radical reactions.These peroxides are reactive and unstable and initiate autocatalytic reaction to generate more and more free radicals which cause a severe membrane, organellar and cellular damage. Damage to cells is reduced when the free radical is captured by vitamin E (a scavenger) embedded in the cell membranes.

(ii) Free radicals cause oxidative modification of proteins.

Oxidation of protein back bone results in protein fragmentation

(iii) Single- strand breaks in DNA occur due to reaction with thymine in nuclei. In short, DNA damage (lesions in DNA) is implicated in malignancy (cancerous transformation of cells) and cell aging in the living body.

5. ELECTRICITY

Static electricity produces no effects in the body. Individuals suffer from discharges of atmospheric electricity or through accidental contact with electrical current (direct or alternating). Lightning has discharges lasting for a short time and causes petechial or linear haemorrhages in the skin and haemorrhages from the nose and mouth. In skin, tissue may be burnt or damaged leaving behind holes or open wounds resembling gun shots injuries.

Direct current acts faster than alternating current which is more dangerous on account of tissues offering less resistance to it than to direct current. Alternating curerent of 100-150 volts produces a strong lethal effect. The harmful influence of current is much greater when a current is applied on moist skin. In producing effects on organisms, the direction of the current is very important. The animal which tolerates the passage of current through its head dies if one of the electrodes is applied to the fore paw and the other to the hind paw.

Duration of the current application is very important. A current of high voltage and great strength is not fatal if it acts for less than 0.1 second. Electrick shock produces burn injury in the skin, convulsions, respiratory paralysis and cardiac arrest.

Three kinds of electrical effects are the following :

(i) Electrolytic

(ii) Electrothermal

(iii) Electromechanical effect

Electrolytic effect

There is a biochemical or electro-mechanical change in the tissues due to electrolysis.

Electromechanical effect

In this, there is conversion of electrical energy into mechanical energy which causes structural damages in the tissues.

Electrothermal effect

In this, the electrical energy is converted into thermal energy resulting into burns. In lightning stroke, there is unconsciousness, burn and death. Rigor mortis develops rapidly in such dead animals. Congestion is noticed in the meninges and lungs. Pulmonary oedema and haemorrhages in other organs also occur and capillary haemorrhage is noticed in central nervous system. There is respiratory paralysis but cardiac activity is maintained. Artificial respiration is advisable to save life in persons exposed to electric shock.

Some important lesions in dead animals from electrical injuries (atmospheric electricity) are as follows :

1. Tree shaped, branching, reddish or bluish streaks on the body.
2. Hairs between the streaks are singed or the skin is scorched.

3. Paralysis of cutaneous vessels.
4. Muscles in intense contraction after a minute of exposure with ultimate death.
5. Presence of deep burns upto even bone at the point of contact of the electric current in the body with singeing or peripheral radiating lacerations. Burns due to electricity are difficult to heal.
6. Pinpoint haemorrhages on the serous membranes of the internal organs.
7. Blood is black or liquefied.

6. THERMAL FACTORS

Thermal effects are produced by heat which can produce burns in small areas due to small hot objects. Whole body can have burn injury due to hot air coming in contract with it. Cold can also lower body temperature with fatal result and local effects can be seen in the form of wounds. (e.g. frost bite)

Local effects of heat

Some increase in temperature can be tolerated by body and leucocytes in body are very active at peak of fever (say, 104^0 F) At 120^0F, the leucocytes of the higher organisms die, coagulate and disintegrate to become irregular or covered with knobs or stalks. These knobs get pinched off and are found in plasma. The red cells become irregular and spherical. At 140^0 F, there is haemolysis of red cells with release of haemoglobin. At 150^0 F, there is coagulation of protein of red cells.

The chief kinds of burns are :

(1) 1st degree burn

In this stage, hyperaemia, swelling, warmth and pain are found in the affected tissues.

(2) 2nd degree burn

There is vesiculation due to accumulation of fluid between dermis and epidermis. Exudative fluid on the base of dermis is rich is fibrin.

(3) 3rd degree burn

In this, necrosis is the main change and dermis is involved. Necrosed tissues have light cherry red colour and may look like brownish yellow leather. Heat and thrombosis cause necrosis in the tissues.

(4) 4th degree burn

Charring or complete destruction of tissue is noticed in the body. Muscles and tendons are exposed or torn due to their contraction or retraction. The local and general effects of burn may cause death. Death may occur during the first three or four days after burning or due to secondary complications.

Effects of burns are as follows :

(i) Death in a few hours due to burn and shock from reflex inhibition of nerve centres.

(ii) Fall of body temperature and difficult respiration.

(iii) Haemolysis.

(iv) Polycythaemia Histamine and histamine like substances produce shock on absorption from the site of burn in the body. In shock, capillaries dilate and get filled with blood.

(v) Thrombosis.

(vi) Hyperaemia of thoracic and abdominal organs.

(vii) Hyperaemia and haemorrhage in the pericardium and gastrointestinal tract.

(viii) Hyperaemia and haemorrhages in adrenal glands.

(ix) Pneumonia, septicaemia and intoxication due to decomposition products of burns.

(x) Extensive burns (1 /4th to 1 /3rd of the whole body surface) may cause death within 24 hours.

The state of hyperthermia or overheating (heat stroke) can occur in an individual when work is done in hot humid air or in warm clothes. It arises from interference with release of excess heat from the body.

The effects of over-heating are as under :

1. Excitation and convulsion.
2. Depression of nervous system (unconciousness).
3. General weakness and depression with emesis and anuria. Body temperature rises to 106-113^0 F.A severe irrepairable damage is done to the tissues of the body.
4. Accelerated respiration and tachycardia, rapid and weak pulse.
5. Paralysis and death from the depressed respiratory centre.
6. Reddened mucous membrane which may be blue with rise in temperature and animals die in convulsions.

The postmortem changes of heat stroke are as under :

(i) Red skin and slow coagulation of blood

(ii) Dilatation of right side of heart and congestion, haemorrhage and oedema of lungs or degenerative changes in the central nervous system of dead individuals.

Effects of cold

The effects depend on the lower limit of the temperature and length of experience. The action of cold gives rise to vascular spasm or sensation of cold and pain. The skin turns pale, and its temperature drops. The blood vessels in the skin dilate due to paralysis of vasomotor nerves and loose tone and get filled with blood. The fluid passes in the injured tissues to produce oedema.

Three stages of frost bite are as follows :

1. Local effects

In frost bite, there is anaemia due to vascular contraction, other changes like active vasodilation, vesicle formation and necrosis follow in the exposed tissues to cold. (i.e. an extreme of temperature)

1. 1st degree frost bite

There is swollen, red, or pale and hot skin.

2. 2nd degree frost bite

It is recognized by vesicle formation with haemorrhagic contents.

3. **3rd degree frost bite**

In this, there is a blue, purple blue or white discoloured skin, Haemorrhagic vesicles may also be seen.

At first, there is a feeling of cold with frost bite which is followed by hyperaemia, Skin, later, shows inflammation and gangrene may follow in the affected tissues of the body.

General effects of cold

1. Hypothermia (Over-cooling)

It is due to drop in external temperature and the individuals become unable to regulate their body temperature. Poor nutrition, inadequate clothing and decreased metabolism help in the development of chills. Exposure to temperature (10-15^0 c) below normal temperature causes chills. Due to chills, somnolence, diminished respiration, fall in blood pressure, unconsciousness and paralysis of nervous centres are noticed. Hypothermia and death may occur at -5^0 C. Microbes enter the dilated blood vessels.

(2) Spasm of the vessels at the site of cooling i.e. local anaema due to contraction of blood vessels.

(3) Sudden dilatation of vessels in the mucosa of air passages.

(4) Development of bronchitis, endocarditis, nephritis and arthritis. A sudden immersion of dog's paws in water at 4^0 C may cause nephritis. There is a drop in their metabolic processes and body temperature.

(5) Fall in phagocytic power of neutrophiles and reduction in production of cytolytic and bacteriolyic substance.

(6) Stiffness of legs and difficult movement. Heart and respi-

ratory centres in brain are depressed and animals die at temperature of 90^0 to 95^0F.

(7) Swelling and redness of the skin.

Freezing causes formation of ice crysalts in the cells and tissues become firmer, more doughy and less elastic. With further drop in temperature, the colloids in the cells coagulate and, thus, an irreversible state is produced and the affected cells die. Blood clots are formed in the vessels in the cases of frost bite.

7. MECHANICAL AND TRAUMATIC FACTORS

Mechanical or traumatic injuries are caused by stone, knife, bullet and sticks and the tissue alterations are almost the same. Tissues are crushed and torn with death of cells and hemorrhage occurs from the damaged vessels in the neighbouring tissues. Servere injury may lead to massive haemorrhage and death.

The pathological effects of the mechanical injury depend upon the following effects :

1. Nature of the injury
2. Nature of force applied
3. State of injury

The effects of mechanical injuries are as under

(i) Local effects.

(ii) General effects

(iii) Occurrence of complications

Kinds of injuries

These are as follows :

1. Sharp injuries

These are injuries caused by sharp objects like knives or swords.

2. Blunt injuries

These are caused by pressure or concussion. Rapid or violent concussion causes crushing or contusion of soft tissues Blunt objects produce such injuries.

3. Penetrating or perforating wounds

These can cause haemopericardium and an excess of blood in the pericardial sac leads to cardiac tamponade, embarrassing the heart. Such wound has a narrow point of entry of the mechanical force as caused by a bullet or nail.

4. Contusion or bruise

There is no break in the integument but the underlying structures are injured with haemorrhage into the surrounding tissues.

Haemorrhage causes redness of the injured areas and these, later, turn into black, blue, green and yellow areas due to disintegration of blood cells or haemoglobin.

5. Abrasion

It is an injury in which the integument is broken.

6. Incision

It means a smooth, long narrow or clean type of wound caused by sharp object in the body. Pointed or sharp objects interrupt the continuity of the tissue to produce incised or punctured wounds.

7. Laceration

It is caused by tearing of the tissues due to blunt objects. Accidents and lightnings produce lacerations.

8 Rupture

It is an injury in which the fibres part to cause stretching of the tissues. A crushing below or excessive distension can cause it and it can be seen in the hollow organs or capsules of liver and spleen etc.

9. Fracture

It refers to an injury of bone, cartilage, tooth, hoof, horn or claw in which the continuity of the hard tissue is broken. In simple fracture, there is an injury in the osseous tissue with no break in the skin. In compound fracture, the skin or integument is also broken.

10. Concussion

A severe jarring injury to the head can cause concussion of the brain.

Jarring causes pressure waves that rebound back and forth in the central nervous system lying in the skull. There is a disturbance of molecular equilibrium.

11. Sprain or strain

This is a kind of injury in which supporting ligaments around the joint are stretched or torn slightly.

12. Laxation or dislocation

It refers to an injury of joint in which the anatomic relation of the bony structures is disturbed and the ligaments supporting the joints are broken. The ligaments and tendons loose the elasticity.

13. Blast injuries

If the mechanical factor is an explosion, the air or gas blast jars the tissue and disturbs the physio-chemical state, usually without involving the coarse anatomic changes. A blast may injure the entire organism or may injure all the organs or tissues.

An injury may cause haemorrhage and haematoma. Haematoma is a collection of effused blood in the tissues. Injuries to capillaries results in petechiae or eochymoses. Blows to the epigastric region, extensive wounds, fracture, or crushing of the tissues may give rise to traumatic shock in body. There is a drop in blood pressure and depression of all the vital functions. Injuries to lungs can cause traumatic haematuria. Traumatic injuries may produce the following changes :-

1. Zenker's degeneration or necrosis.

2. Inflammation

Histamine causes dilatation of capillaries enabling them to hold more blood than usual and drop in blood pressure follows.

3. Primary shock (Neurogenic shock)

It results from the vasodiatory effects of histamine and there is unconsciousness in the affected individual.

4. Secondary shock

It may occur in 2 to 24 hours after trauma and results from haemo-concentration. Reduction in blood volume, fall in blood pressure and inadequate circulation lead to hypoxemia and damage to the capillary walls.

5. Fat embolism due to fracture of the bone and entry of fat into circulation.
6. Neoplastic growth due to traumatic factor.
7. Bacterial infection as a complication at traumatic site.

Acoustic waves

These waves cause traumatic injuries. A shot (e.g. sound of the gun fire) or locomotive whistle may produce a pain sensation and can cause damage to the ear drum or internal ear. A sharp electric bell gives rise to disturbances in the central nervous system. Intense noises can cause neuropsychic disorders like insomnia, headache and deafness. Ultrasonic waves can cause heat in the affected tissues.

8. NUTRITIONAL FACTORS

Ingestion of insufficient or excessive amounts of nutrients and unsuitable foods causes nutritional disturbances. Nutrients must be satisfactory both quantitatively and qualitatively. The amounts of nutrients ingested depends upon individual predisposition, work performed and external factors (e.g. climate). Proteins, lipids and carbohydrates are the main dietary recquirements. An extreme protein diet leads to metabolic disorders like gout. Deficiency or absence of essential nutrients in the diet causes diseases in the individual's body. Diet rich in proteins gives more resistance to diseases than the diets rich in carbohydrates. Fat deficient diets produce atrophy of glandular tissue (sexual glands), osteoprosis and atrophy of bones and Zenker's degeneration of muscles. Feeding of a lare amount of cholesterol causes deposition of lipids into intima of arteries (atheromatosis).Excessive amount of egg yolk in diet produces arteriosclerosis in humans. Starvation leads to hunger oedema which is recongnised by generalized atrophy, oedema and haemosiderosis. A diet rich in cellulose but deficient in fat and protein with production of about 1,000 calories per day will lead to hunger oedema.

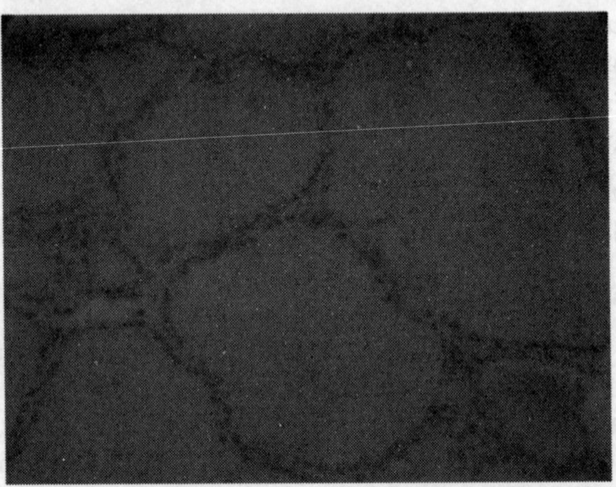

Fig-1 Colloid goiter, foal, Note the deposits of the colloid distending the thyroid follicles. H & E x 400

Absence or deficiency of minerals causes diseases in the body of an organism. Iron is used by haematopotetic tissue to produce haemoglobin which gets incorporated into the red cells. Iron deficiency causes lack of haemoglobin (i.e. a state of anaemia). Bone formation is badly affected due to lack of calcium, particularly in the young. A calcium free diet causes rickets, malacia or osteitisfibrosa. Phosporus deficiency causes osteomalacia in adult cattle. In nervous system, calcium decreases irritability but potassium and sodium normally increase it. Potassium, magnesium and phosphates predominate in the cells whereas; sodium, calcium and chloride ions do so outside the cells. Potassium suppresses leucocytosis but calcium enhances it. High level of intracellular calcium damages the cells. Change in acid base balance is caused by an excess of potassium or calcium. A change in ion balance of potassium and calcium in cells is caused by inflammation, Excessive intake of sugar produces acidosis in rabbits, atrophy of the bone and fibrousosteodystrophy of the bones. An inadequate supply of vitamin D and parathyroid insufficiency causes rickets in young animals. There should be proper assimilation of calcium and phosphorus along with availability of vitamin D to check rickets in animals. In rickets, the bones become soft, pliable and deformed with enlarged epiphyses of the long bones.

Iron deficiency in the diet leads to poor growth, anemia, loss of weight and depraved appetite in animals. Iron is contained in iron-prophyrin compounds, like hemoglobin, myoglobin, cytochromes and enzymes like oxidases and peroxidases. The active form of the iron is its divalent ferrous form. Copper deficiency gives tise to anaemia, diarrhoea and loss of normal hair pigment. Iodine is a free ion or bound organic compound in tissues and is essential for thyroid function. It is found in the hormone thyroxin which is essential for physical and mental development and also for proper metabolic rate. Deficiency of iodine causes goiter, cretinism, disturbances in growth and development and poor viability of foetus and the new born. (Fig-1) An animal must have adequate quantity of water. Water is being lost though the skin, lung, faeces and urine. If there is no proper hydration of the blood and tissue fluids, there is a feeling of thirst. But too much water intake causes increase in the volume of blood and work to be performed by the heart. When there is lack of water, animal refuses to eat and develop the state of inanition or debility. Longer periods of starvation have effects on the body. Dogs may starve for up to about 2 months. Humans die after 1-2 weeks of starvation but some people accustom themselves to leave without food for 3 to 4 weeks, in case, water is available during this period. Starvation causes decline in body temperature, pulse and respiratory rates.

Prolonged starvation causes intestinal haemorrhages and an atrophy of tissues and organs sets in the body. When there is a loss of about 48% of the body weight, an animal dies. Organs like spleen, liver and skeletal muscles show wrinkle (e.g. spleen) are reduced in the bulk of tissues (i.e. smaller than normal). Brain shows decline in weight (about 4%). Adrenals and bones show small losses in weights or sizes. Dehydration causes increase in red cell count (polycythemia) and loss of glycogen (hypoglycemia). Excessive intake of nutrients is also very bad. Too much fat in the diet produces obesity or adiposity in the body of an individual. Insufficiency of oxygen causes serious effects which can result from insufficient oxygen in the air or blockage of the respiratory passages. Animals die when there is a drop in oxygen content of 21% in the air to 2 or 3 percent. Blockage in the respiratory passage is caused by fetal fluids or enlarged thyroid. In hanging, asphyxia is produced in the body. There is a discontinued access of air to the lungs as noticed in hanging, strangulation, drowning, pulmonary oedema and obstruction due to foreign bodies in the respiratory tract.

A bluish red color (cyanosis) in the skin or mucous membranes arises from a drop of oxygen content, convulsions, vascular spasm, mydriasis and bradycardia are produced in oxygen deficiency. The body fails to react and coma and death follow. Moreover, the organs are congested and ecchymoses are seen in the pleura and pericardium. When the level of **carbon-dioxide** is high, blood fails to clot. Oxygen deficiency leads to interference with tissue respiration resulting in degenerative changes like cellular swelling, fatty changes and necrosis in organs and the cause of death is **asphyxia** as noticed in all diseases of organisms.

9. ATMOSPHERIC PRESSURE

Both the lowered and elevated atmospheric pressures produce changes in the body of an individual. Changes in the body could be seen at an altitude of 4000 m. There is a lowered partial oxygen pressure in the inhaled air at high altitudes. Climbers on mountains suffer from sickness (mountain sickness). At high altitudes, air pilots suffer from altitude sickness. Lowered atmospheric pressure produces fatigue, dizziness, headache, dyspnoea, tachycardia and metabolic disturbances. Prolonged staying at an altitude of 7000-8000 meters usually results in unconsciousness and even death. These symptoms arise from an oxygen deficiency or state of anoxia, hypoxia and asphyxia etc.

Rarefied air produces following effects in an organism :

1. Reflex acceleration of respiration or pulmonary ventilation.
2. Acceleration of blood flow and contraction of the spleen with increase in number of red cells.

Elevated atmospheric pressure causes pathologic changes in the body of an individual as seen in the cases of persons engaged in deep sea diving or cassion work.

There is an increased partial pressure of oxygen in the inhaled air. An individual under atmospheric pressure of 2-3 atmospheres i.e. at the depth of 10-20 m under water shows slowing down of pulse and respiration, rise in blood pressure, overfilling of internal organs, marked inhibition of central nervous system, general convulsions and unconsciousness. If there is a rapid return of an individual from elevated atmospheric pressure to normal atmospheric pressure (i.e. the state of sudden decompression), gases dissolved in the blood at elevated atmospheric pressure are liberated in the form of bubbles which block the small blood vessels like capillaries or venules etc. (i.e. a state of air embolism) The disease produced is called caission disease which shows pain in the muscles, joints, itching on the body, respiratory and circulatory disturbances. The dissolved gases include particularly excess of nitrogen and the condition produced is also called araemia or azotaemia. Such gases dissolved in the tissues cause distortion or distension of the cells.

To sum up, the main injurious causes of diseases in animals are as follows :

(1) Oxygenation deprivation

Its two causes are :

(a) Hypoxia (inadequate blood in the oxygenated form)

It produces both the reversible and irreversible degenerative changes in the damaged tissues. Cardiovascular and respiratory diseases result in inadequate oxygenation of blood in the organs of body. Death occurs in diseases in organism in the state of asphyxia characeterised by insufficient delivery of oxygen to tissues as seen in the cases of drowning, obstruction in the respiratory tract and pulmonary edema and high level of tissues carbon dioxide content.

(b) Ischemia

It arises from the stoppage or reduced supply of blood to a part of an organ because of decreased venous drainage, embolism and thrombosis. Both reversible and irreversible degenerative changes are noticed in the ischemic organs.

(2) Free radicals like super oxide anion, hydrogen peroxide and hydroxyl ions.

(3) Physical agents e.g. pathogenic effects noticed in the extremes of temperature as exemplified by burns and frost bite.

(4) Chemical agents e.g. arsenic, carbon tetrachloride, mercury salt and cyanides.

(5) Infectious agents like bacteria, viruses and parasites.

(6) Immunologic reaction e.g. anaphylaxis.

(7) Genetic derangements.

Genetic defects cause cellular injury. Alterations at the level of DNA cause cell damage and metabolic errors arise from enzymatic abnormalities.

(8) Nutritional imbalances

Protein deficiency causes deaths of animals. Vitamin A deficiency enhances susceptibility of chickens to infections of the respiratory tract. Excess of lipids in the food predisposes humans to atherosclerosis.

PATHOLOGICAL CHANGES

These are divided into two types, namely, (i) Retrograde changes and (ii) Progressive changes. Degeneration (L = to degenerate) and infiltration are reterograde or retrogressive changes which may be reversible and irreversible in nature.

When cells or tissues in the living organisms are damaged due to exogenous harmful agents, pathological changes occurring in the cells vary according to the severity or nature of the irritants, duration of the irritation and sensitivity of the affected cells. Mild irritants acting over some period in the cells stimulate them to proliferate for defensive and reparative purposes which are called progressive changes. Hyperplasia is an example of the progressive change. But irritants of severe intensity act as poisons cause disturbances in the cell metabolism and retrograde changes like cellular swelling and fatty changes in the affected cells. These changes may be reversible and intermediate stages in the whole dying or deteriorating processes of cells, and with removal of the causes, the affected cells recover. Strong irritants produce quick irreversible changes (e.g. outright death or necrosis) in the cells. Death of the cells in the living body of an organism is called necrosis. Dead or degenerated tissues, later, may be infiltrated with calcium salts (e.g. calcareous infiltration). Thus, retrogressive changes include degenerations and infiltrations. Degenerations refer to deterioration of tissues or organs with loss in both vitality and physiological activity. These changes result from lack of proper oxygenation (anoxia) caused by bacteria, virus, parasites, chemicals, failure of circulation and anemia etc. Cellular metabolism is disturbed and intracellular enzymes do not function properly. Degenerations cause disintegration of the cells and tissues which, later, disappear from the scene or site of an injury followed by reparative changes (e.g. replacement fibrosis or fibroplasia).

Infiltration means deposition (entry) of some foreign substances into the cells from without. These abnormal accumulations of some substances in the cells are usually not found in such excessive or abnormal amounts. Entry of fat, glycogen and carbon particles into the cells of the body constitutes this state of infiltration.

Some degenerative and infiltrative changes in diseases are as follows :

Degenerations	Infiltrations
1. Cellular swelling	1. Fatty infiltration(a fatty change)
2. Cellular edema	2. Amyloid infiltration
3. Fatty degeneration (a fatty change)	3. Glycogenic infiltration
4. Necrosis	4. Pathological calcification
5. Gangrene	5. Pathological pigmentation
6. Hyaline degeneration	6. Uratic or gouty infiltration(Gout)
7. Mucoid degeneration	
8. Atrophy.	

The main types of cellular injuries are :

i. Hypoxic and ischemic injuries.

ii. Injuries caused by free radicals.

iii. Injuries induced by bacterial toxins, viral proteins and chemical poisons and so on and so forth.

The chief factors influencing reversible and irreversible changes in the body are the following :

i. Types of injuries or stimuli

ii. Duration of activity of the irritant in the affected cells

iii. Severities of the irritants

iv. Susceptibility of tissue, nature of tissues and adaptability of cells and tissues to noxious influences

Some main biochemical effects of injurious agents are as follows:

i. Injuries affect the integrity of cell membrane,aerobic respiration (i.e.mitochondrial oxidative phosphorylation) and production of adenosine triphosphate–ATP), protein synthesis and integrity of the genetic mechanism of the cells. The energy dependent membrane transport mechanism called sodium pump which maintains the ionic and fluid balance of the cells is also greatly disrupted by hormonal or injurious factors.

ii. Deranged biochemical mechanism of the cell produces reversible changes within a few minutes. There is an *earlier development* of ultrastructural changes than the appearance of light microscopic morphologic changes. In total ischemia, the cardiac muscles show light microscopic changes within ten to twelve hours. Microscopic morphologic changes of reversible and irreversible types develop in dead cells of the living body with the passage of time.

iii. ATP depletion

There is depletion of ATP and ATP synthesis as a result of ischemic and toxic injury.

iv. Generation of oxygen derived free radicals. (reactive oxygen species)

Some small amounts of oxygen derived free radicals are produced as byproducts during mitochondrial respiration in cells. Free radicals damage to lipids, nucleic acid and proteins etc. are noticed.

An imbalance between free radical generation and radical scavenging process leads to an oxidative stress. This stress is also associated with cell injury in the body.

v. Disturbance in calcium homeostasis or increased intracellular calcium.

Increased cell membrane permeability is enhanced by high level of intracellular calcium (normal cytosolic calcium less than 01.m mol and extracellular calcium content equal to 1.3 m mol in cells and tissues).

The main effects of intracellular calcium are as under:

i. Activation of the enzyme phospholipase A_2

It produces membrane damage in cells.

ii. Activation of protease

Breakdown of cell membrane cytoskeleton protein is caused by proteases.

iii. Cell death because of an increased intracellular calcium.

iv. Defects in cell membrane permeability

Cells or tissue loose the ability of selective permeability due to activated phospholipase A_2 by enhanced intracellular calcium concentration. Defects in plasma membrane and cell membrane are caused by injurious factors like bacterial toxin, viral protein and perforins released by cytolytic lymphocytes.

v. Irreversible damage to mitochondria

Toxins and hypoxia cause lethal hit to mitochondria in the cells. Increased cytosolic calcium activates the enzymes like ATPase, phospholipase, protease and endonuclease.

The enzymes are given below :

Enzymes	Pathologic effects
1. AT Pase	1. Decrease in ATP levels in cells
2. Phospholipase A2	2. Decrease in phospholipids in membranes.
3. Proteases	3. Disruption of membrane and cytoskeleton proteins Apoptotic death of cells and mitochondrial of permeability transition (MPT) detected in the inner mitochondrial membrane.
	4. Leakage of Cytochrome C in cytosol arising from mitochondrial damage

In short, membrane damage in cells arises from high level of cytosolic calcium Ca^{2+} causing loss of phospholipids, break down of lipids and cytoskeletal structures i.e. microtubles and thin acting filaments and intermediate filaments etc.

CELLULAR SWELLING

The terms like cloudy swelling, acute cellular swelling, parenchymatous degeneration, aluminous degeneration and granular degeneration are no longer in common usage.

It is an earliest but mildest and commonest of almost all the degenerative changes and characterized by cloudy and swollen appearance of the affected organs. The term aluminous degeneration is used signifying the appearance of abluminous or portenious material in the cells of the parenchymatous organs which are found to be most commonly affected. This basic reversible or the first degenerative change arises from injury to the highly specialized parenchymatous cells of organs like liver, kidneys and heart etc. A biochemical lesion develops in affected cells of such organs.

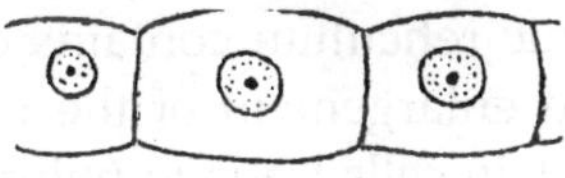

Fig-2 Normal cells

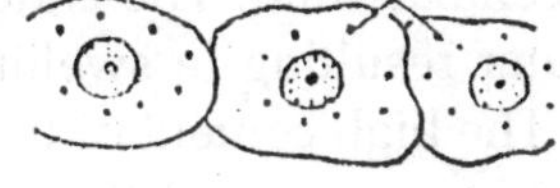

Fig- 3 Cellular swelling

Causes

1. Acute febrile or toxemic diseases. e.g. septicemia and pneumonia.
2. Poisonous substances e.g. phosphorus, chloroform, carbon tetrachloride, snake venom and arsenic etc.
3. Toxins or harmful decomposition products, on account of diseases or severe injury e.g. burns or infections. Plasma proteins enter the injured cells.
4. Circulatory disturbances arising from cardiac stenosis or incompetence, valvular or pulmonary (fibrosis or severe emphysema) diseases. Injured cardiac myocytes due to anoxia or hypoxia show accumulation of albumin, globulin and fibrinogen.
5. Disturbances in active transport of water and cell respiration and cell metabolism.

Temporary rise in cell osmotic pressure and tissue hydration are found. Sodium ions are not extruded from cells and these ions enter the cells with chloride ions. Such changes are responsible for cellular edema in the affected cells.

Failure of the Na+ and K+ pump of active transport system (in other words, the Na+ K+ AT pase transport system) causes accumulation or entry of Na+, Ca+ and water inside the cells.

This leads to cellular and tissue hydration. The accumulating water and granules (proteins) make the cells swollen and vacuolated under light microscopy. The affected cells may, later, break up into pieces. Ca++ (cal ions) stimulate endogenous a phospholipase which causes breakdown of cell membrane. The cell membrane becomes abnormally permeable or leaky causing high concentration of Na+, Ca++ and water in the cells. The intracellular K+ becomes less than the normal concentrations of Na +. The intracellular K + and magnesium escape into the extracellular fluid. The endoplasmic reticulum contains excess of water resulting in swelling and enlargement of the injured cells. The high content of the Ca ++ in cells leads to point of no return i.e. necrosis of the tissues.

To sum up, Na+ and K+ stimulated AT Pase system is marked by an energy dependent extrusion of sodium (Na+) out of the cells and its replacement by K+ to maintain its high concentration in the cytosol. This active transport system is greatly disturbed in cellular swelling leading to influx of sodium, calcium and water in the cytosol because of damaged active transport system.

6. Anemic conditions

Gross appearance

1. Normal transparent appearance of the cell cytoplasm is lost with opacity of the cells due to granules of proteins.
2. Organs are softer than normal and swollen with boiled appearance.
3. Bulging of tissue on the cut surface of the organ.
4. Normal markings of the organs are blurred, for example, the affected liver does not show distinct lobulations.

Microscopic appearance

1. Cells are swollen and normal transparent appearance of the cell cytoplasm (Fig-1) is lost (Fig.2).
2. Cell borders are indistinct and the cytoplasm is granular. The granules can be cleared with weak acetic acid or alkalies indicating them to be proteins in nature.

3. The nuclei may be enlarged and stain lightly due to decrease in the content of deoxyribonucleic acid (DNA).
4. Parenchymatous degeneration is almost indistinguishable from postmortem autolysis.
5. In the affected kidneys, the cells lining the convoluted tubules or ascending limbs of Henley's loops are swollen and granular and even obstruct the tubules. Pallor of the liver is due to compression of its sinusoids by swollen liver cells. Cardiac muscle cells are swollen and show no longer their characteristic striations.

By electron microscopic, the ultrastructural changes are :

i. Blebbing, blunting and distortion of microvilli, creation of myeline figures and loosening of intercellular attachment.
ii. Swelling, refraction presence of small phospholipids and amorphous densities in the mitochondria.
iii. Dilation of the endoplasmic reticulum and detachment and disaggregation of polysomes.
iv. Some nuclear alterations with disaggregation of granular and fibrillar elements.

All these ultrastructural changes are reversible.

Results

1. Affected cells revert to normalcy with removal of the cause. In other words, all the cytoplasmic or slight nuclear changes are reversible.
2. Fatty changes and necrosis may follow cellular swelling.
3. The cellular swelling may advance to the next stage called hydropic degeneration or vacular degeneration.
4. Blisters or ulcers may follow cellular swelling.
5. Loss of ribosomes causes decrease is the cellular pH and protein synthesis.

Hydropic or Dropsical degeneration

It looks like cellular swelling because of its close alliance to it. The term oncosis refers to a prelethal reversible change marked by swelling of cells (oncosis, Greek for swelling). An excessive accumulation of fluid within the cells is called hydropic degeneration. Disturbances of active transport of cells cause cellular edema because of injury to cells.

Gross appearance

There is no marked change in the injured cells

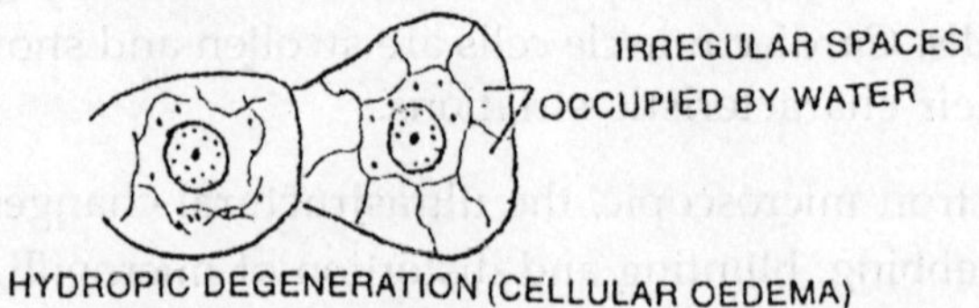

Fig. 4

Miscroscopic appearance

1. Cells are enlarged up to their bursting points with presence of large, clear spaces or vacuoles containing clear fluid. (Fig-4)
2. Cells are reduced to a mere shell and details of the cellular structures are not visible.
3. Skin, liver and kidneys show swollen vacuolated cells.
4. By EM, swelling of the mitochondria and rough endoplasmic reticulum is noticed.

FATTY CHANGES

Fatty changes refer to appearance of fat in cells or tissues due to various factors causing diseases. Fatty changes can be divided into two types known as (1) fatty degeneration and (2) fatty infiltration. A non-committal term like fatty metamorphosis or fatty change (a change in parenchymal cells characterized by the appearance of fat droplets) includes both fatty degeneration and fatty infiltration. Fat exists in the cells in an intimate combination with the cell protoplasm and such fat can be rendered visible or stainable under some pathological conditions. Normally, fat is also stored in the subcutaneous tissue and omentum etc. in the visible form called visible fat.

Fatty degeneration

It refers to the appearance of fat in a visible or stainable form in the parenchymatous cells of the organs which do not reveal stainable fat in health. It is found in the organs like liver, kidney and heart. Fat may also be apparent in capillary endothelium or connective tissue cells. Fatty change in organs like kidney and heart seems to arise due to the process of fatty infiltration. Fat also appears in liver from fatty infiltration in some pathological conditions owing to transported fat absorbed from the gut.

Causes

1. Deficient oxidation or nutrition. Anaemia, starvation and malnutrition cause fatty changes in organs of the body. The fat appears in the liver cells due to interference with the blood supply.

2. Poisons or toxins (i.e. toxic influences)

Phosphorus, chloroform, and arsenic etc. cause fatty degeneration through interference with proper oxygenation. Bacterial toxins in febrile or septicemic conditions also produce fatty degeneration.

3. Parasites e.g. ascariasis in puppies and hook worm disease.
4. Fatty changes may also arise from interference with transport and metabolism of fat.

Excessive release of free fatty acids from adipose tissue can cause this condition. Organs like liver, heart and kidneys fail to utilize the increased delivery of free fatty acids and these are deposited in them as neutral fat. Starvation also follows this mechanism for fatty changes.

2. Decreased utilization or oxidation of fatty acids happens in cases of bacterial intoxications (e.g. diphtheria toxins in the body).

3. Lipotrope deficiency / metabolic injury.

When there is deficiency of methionine or chlorine, there is a decrease in phospholipid synthesis. Fat may accumulate excessively in the cells which rupture with a subsequent replacement by scar.

4. In failure of protein synthesis lipids of liver are excreted in the form of lipoproteins. Protein synthesis is disturbed due to poisoning by carbon tetrachloride or yellow phosphorus, in the ribosomes of the endoplasmic reticulum. The result is an accumulation of neutral fat in the liver cells.

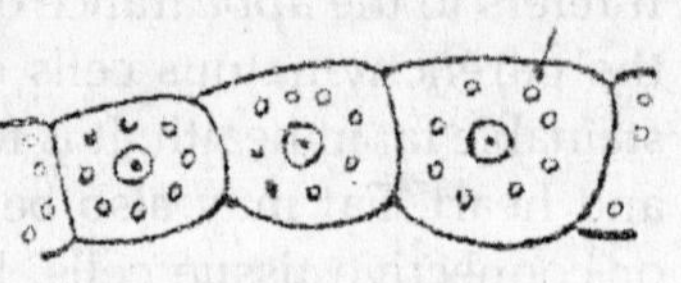

FATTY CHANGES IN LIVER CELLS

Fig. 5

GROSS APPEARANCE

Liver

1. Pale, yellow or clay colour in appearance.
2. Liver may be enlarged.
3. Swollen and rounded edges.
4. Soft or friable liver and greasy to touch with fingers. Fat droplets can be seen on the edge of knife, while cutting the liver during autopsy.

Kidneys

1. Pale or sometimes enlarged. Cortex may look pale on cutting the kidneys into two halves.

Heart

1. Pale, flabby or greasy to the touch in febrile and toxic conditions due to diffuse distribution of fat.
2. An even distribution of fat in the subendocardium gives appearance of striped or flecked markings. Such peculiar appearance of the interior of heart due to fat deposits under the endocardium gives a characterstic name to the heart, i.e. thrush breast or tabby cat heart. There may be orange yellow or brown mottling of the heart, especially, on the inner portion of left ventricle in human beings.

Microscopic appearance

Sudan III, scarlet red, osmic acid or nile blue sulphate are used to stain fat in the cells. Neutral fats and fatty acids stain red with Sudan III or scarlet red in frozen actions. Osmic acid stains triolein and oleic acid black. Fats and lipids are seen as clear, unstained round spaces in cytoplasm of the affected parenchymatous organs e.g. liver following the staining of sections of liver by haematoxylin and eosin technique (Fig-5). Affected liver cells may show one or more large or small droplets of flat. Fat can be seen in the cells lining the convoluted tubules of the cortex and the loops of Henle. Fat droplets may be numerous and small. Fat globules are seen in longitudinal rows in the cardiac muscle fibers.

Fatty infiltration

This refers to an excessive accumulation of fat in the normal storage depot of the body in various pathological states. It is found in the condition of adiposity called fatty infiltration.

Causes

1. Excessive consumption of fat and carbohydrates i.e. unbalanced diet or ration
2. Too little exercise
3. Deficiency of pituitary secretion

PATHOLOGICAL CHANGES

Heart

Fat is found beneath the epicardium along the coronary grooves. It can enter into heart making a covering of the coronary artery. The heart may be encased in a bag of fat. There is an interference with cardiac movements and heart may dilate or rupture due to infringement. Microscopically, fat is always found outside the muscle cells i.e. in the intermuscular connective tissue.

Liver

The liver stores fat excessively. The infiltrated fat is seen in the peripheral liver cells of the hepatic lobules and fatty infiltration may extend to intermediate or central zone of the lobules. Fat deposition is intracellular or may form large fat globules. The nuclei of the liver cells are flattened or may be pushed to the sides of the cells.

Effects

1. Affected organs may rupture leading to death.
2. Fatty changes may occur in the cells of an organ in association with necrosis.
3. Functional disturbance is seen in case of heart loaded with fat.

Necrosis

The term necrosis (Gr = deadness) refers to the death of tissue or a part of the living body of an organism. Another term necrobiosis (Gr = dead body + life) is often used to indicate physiological death of cells in the living body. The continuous death of the superficial cells of the skin (e.g. epidermis) is an example of necrobiosis.

Causes

The causes of necrosis are as follows :

1. Interference with nutrition. An embolus, thrombus, contraction of the blood vessels or pressure on the blood vessels from outside (e.g. tumor and ligatures etc.) obstructs the vessel which leads to local death of cells (infarction).
2. Physical agents like extreme heat and cold, X-rays irradiation and prolonged pressure cause death of the cells. A badly fitting harness causes bedsores in horses.
3. Pathogenic animate factors e.g. bacteria, and parasites etc. *Fusiformis necrophorus* and *Mycobacterium tuberculosis* cause necrotic changes in the affected tissues.
4. Poisons

Carbolic acid, caustics or mineral acids damage or straight away kill the cells. Ergot causes death of the tissue by indirectly influencing the blood supply due to extreme narrowing of the lumina of the blood vessels.

5 Enzymes

Lipase released from pancreas can cause death of fat cells as seen commonly in the peritoneal fat.

6. Nerve lesions

Plantar neurectomy causes necrosis of the foot in horses. Necrosis of the cells depends upon factors like intensity or concentration of the irritant and susceptibility of the tissue and blood supply to the part. Stoppage of blood supply in central nervous system causes liquefaction necrosis of the affected part. Skin does not suffer excessively from arrest of circulation for a few minutes.

Kinds of Necrosis

1. Coagulation necrosis

It arises from an arteriolar occlusion in organs like kidneys and spleen which have poor anastomosis with more or less end arteries.

Gross changes

Swollen, solid or firm dead tissues with a dull white or yellow discoloration. Lymph is absorbed by the dying cells and undergoes coagulation to impart firmness to the dead tissue.

Fig-6 Hepatic necrosis. Note the pyknotic fragmented nuclei and lack of walls in the cells H & E x 400

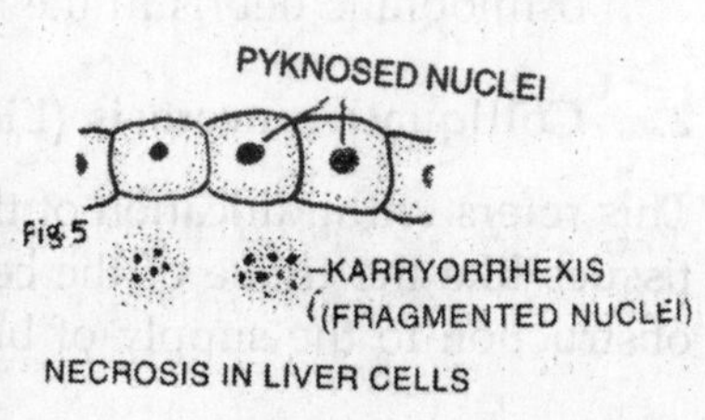

Fig. 7

Microscopic appearance

In the H & E stained sections of tissues, the cellular details of the dead cells are lost but the structural outline is still discernible. The dead tissues may be structureless or homogeneous and staining characteristics of the cells are lost. Nuclear changes are visible and these are as follows :

1. **Pyknosis** (Fig. Gr. = condensation). It is characterized by hyperchromatism and condensation of the nuclei (Fig.6).
2. **Karyorrhexis** (Gr. = not to split) In this, nuclear membrane breaks into parts dispersing the chromatin in the form of irregular, deeply staining and intensely basophilic granules (Fig. 7).
3. **Karyolysis** (Gr. = nut + loosening) or Chromatolysis (Gr. = colour loosening) in which the nucleus enlarges, stains lightly or becomes watery and disappear from the view. In karyolysis, the nucleus is lost in the cytoplasm. The whole tissue is fused into a homogeneous mass which takes pink colour with eosin.

By electron microscopy, the ultrastuctural changes are as follows:

1. Discontinuities in plasma and organellar membranes and dilation of mitochondria and presence of large amorphous densities
2. Presence of intracytoplasmic myelin figures and amorphous osmiophilic debris in the cytoplasm.

2. **Colliquative necrosis (Liquefaction necrosis)**

This refers to fluidification of the involved tissue and occurs in tissues like the tissue of the central nervous system following obstruction to the supply of blood.

Gross appearance

The centre of the lesion is replaced with fluid and the periphery of the lesion consists of soft grey friable tissue.

Microscopic appearance

Presence of empty spaces due to liquefaction of the affected tissues is viewed in the nervous tissue.

3. Fat necrosis

Lipase released from pancreas causes necrosis of the fat. Fat is split into fatty acids and glycerol. The fatty acids are seen as crystals in the tissues. Dead fat is firmer in nature and dull white or yellow in colour.

Microscopic appearance

In ordinary haematoxylin and eosin tissue sections of the affected tissues, the fat cells stain with eosin and have a structureless or opaque appearance. Calcium soap is precipitated to form a granular bluish straining material between healthy and dead adipose tissue. A zone of inflammation can be found beyond the dead zone behaving as an irritant

4. Caseation necrosis

The term caseation (L. = cheese) necrosis refers to a kind of cell death in which the tissues are converted into a cheesy structureless substance. The cheesy material consists of numerous fine fat droplets and protein granules. It occurs in tuberculosis in man and animals. Caseations necrosis is found in abscesses and old infarcts.

Gross appearance

The affected tissue is dry and crumbly and slightly greasy. It can be broken into granular fragments and may look grayish white in colour.

Microscopic appearance

There is destruction of the tissue architecture and fine amorphous eonsinophilic granules are seen in place of the cells in the haematoxylin and eosin sections. Calcium salts are present in the caseous material and take purplish colour from the basic haematoxylin.

5. Zenker's necrosis (Zenker's degeneration)

It is noticed in striated muscles and occurs in the form of coagulation of proteins of the sarcoplasm. Bacterial toxins, selenium and deficiency of vitamin E can produce such necrosis as seen in white muscle disease of ews . Grossly, the affected muscle is pale or white and may be somewhat swollen.

Microscopically, the muscle fibres are swollen, homogenous and hyaline in nature. The sarcoplasm is highly red i.e. highly acidophilic in haematoxylin and eosin sections of the tissue. The nuclei of the muscle cells are dark, small and the myofibrils in the muscle fibers are not visible.

Results of necrosis

1. Growth of putrefactive bacteria in the dead tissue gives rise to gangrene.
2. Dead tissue may be cast off as a slough.
3. Leucocytes invade the tissue and produce enzymes which liquefy it. The liquefied material may be absorbed. There can be either regenerative changes in organs like liver or replacement of the dead tissue by fibrous tissue. Infarcts of dead tissue can be encapsulated by a fibrous tissue wall. The dead tissue can give rise to sequestration as seen in the lungs affected with contagious bovine pleuropneumonia. In lungs, a sequestrum refers to dead tissue encapsulated in fibrous tissue growth. Small sequestrum is absorbed in the lungs and probably arises from thrombosis.
4. Dead tissue may be calcified or ossified.

Apoptosis (Gr. = falling off)

It denotes the regulated mode of cell death by genes (i.e. a death controlled by genes or a genetically programmed cell death). It is considered to be a kind of death differing from the death of the cells in living body (called necrosis) when an organ is exposed to harmful agents in environment. This process of death of cells eliminates unwanted cells of the body involving the roles of certain genes. In short, it is an internal genetically programmed cell death in the body. There is a some suicide programme of unwanted cells as shown by certain cells of tissues during embryogenesis. Developmental involution is marked by programmed destruction of cells before delivery of the foetus. Many physiologic, adaptive and pathologic events reveal this process of death called apoptosis. Hormonal dependent involution is noticed in adults. e.g. regression of the lactating breast after weaning and prostatatic atrophy after castration.

The chief types of apoptosis are as follows:

1. Cells deletion in proliferating cells e.g. loss of intestinal crypt epithelial cells.
2. Death of immune cells.

   Both the B-and T lymphocytes die after depletion of certain cytokines.
3. Cell death induced by cytotoxic T cells.
4. Involution of thymus, atrophy of certain organs or remnants of certain structures and disappearance of digits on the limbs.

By electron microscopy (EM), the main ultrastructural changes in apoptosis are :

1. Cells shrinkage

The cells are smaller than normal with dense cytoplasm. The organelles in such cells are densely packed.

2. Chromatin condensation

Aggregates of chromatin forming dense masses of different shapes and sizes are noticed under the nuclear membrane. Nuclear fragments may be seen in the cytosol.

3. Formation of cytoplasmic blebs (or buds) and apoptotic bodies

Surface blebbing is shown by apoptotic cells and fragmentation of membrane in apoptotic bodies. The bodies are composed of cytoplasm and tightly packed organelles (with or without nuclear fragments). Round or oval apoptotic bodies reveal highly eosinophilic cytoplasm with dense chromatin fragments.

4. **Phagocytosis of apoptotic bodies**

Macrophages and parenchymal cells engulf such bodies. These bodies are destroyed by lysosomal enzymes. Proteases disrupt cytoskeleton and cause cells shrinkage. Intact or permeable plasma membrane may be noticed.

5. Lack of inflammatory reaction is not invoked by apoptotic cells around themselves. Protein hydrolysis involving caspases and DNA breakdown are noticed in apoptosis. Necrosis differs from apoptosis because of presence of an inflammatory zone around necrotic tissues which act like irritants.

Gangrene

The term gangrene refers to death of tissue with addition or multiplication of putrefactive or saprophytic organisms in them. Both the exposed and internal portions of the body can be affected by gangrene.

Causes

1. Interference with circulation e.g. gangrene in the extremities.
2. All the causes of necrosis plus the exposure of the dead tissue to saprophytic bacteria producing gangrene.
3. Toxic products of bacteria as seen in cases of gangrene of lungs and udder.
4. Administration of irritant or medicines in the lungs by mistake. Irritants kill the pulmonary tissue and set the stage for growth of the putrefactive bacteria i.e. necrosis is caused by different kinds of factors whereas gangrenous changes are produced by invasion of already dead tissue by saprophytes.

Gangrenes can be divided into primary and secondary gangrenes. There is a third kind of gangrene called gas gangrene which is marked by presence of gas in the affected tissues. The causes of both necrosis and putrefactive changes are the same in the primary type of gangrene whereas in the secondary gangrene, both conditions (i.e. necrosis and putrefaction) are caused by two separate sets of causes. Gangrenes can also be expressed as given below :-

(i) Dry gangrene

The tissues containing little fluid favouring evaporation suffer from dry gangrene. Ears, limbs and skin etc. are affected with dry gangrene and are always in contact with atmosphere. Such tissue conditions are not strongly conducive to bacterial growth. The gangrenous changes slowly progress in the body as seen in Dengnala disease.

Grossly, the affected parts are dry, cold and have a shrivelled. diffused appearance. Hemoglobin can cause pigmentation of affected tissues. Healthy and diseased tissues are easily demarcated by inflammatory zone of reaction between the two tissues. Gangrenous tissues are separated from healthy zone by protective inflammatory changes. Microscopically, the necrotic changes are found along with putrefactive bacteria in ordinary haematoxylin and eosin sections.

(ii) Moist gangrene

In this, there is at first death of tissues which are later invaded by the putrefactive bacteria. The internal organs like lungs and intestines which do not favour evaporation are affected by moist gangrene. It is seen in the tissues filled with blood or fluid.

Intestinal gangrene arises from interference with blood supply to intestine. Necrosis (ischemic) arises from embolism and thrombosis of blood vessels nourishing the intestine. Saprophytes or putrefactive bacteria grow in the dead tissues to convert it into moist gangrene of the intestine. Moist gangrene also develops in the lungs following faulty drenching of drugs or milk. Pulmonary parenchyma is destroyed by irritants (drugs) and putrefactive bacteria grow in the necrosed lung parenchyma.

Gross changes

1. Affected part is swollen, soft, pulpy or usually dark or greenish black in colour.

2. Foul smell in the dead tissues.
3. No heat and insensivility to touch or no pain in the dead tissue.
4. Emphysematous appearance of dead tissue as seen in *Cl. chauvoei.*
5. No line of demarcation between dead and healthy tissue in organs.

Microscopic appearance

1. Presence of rod shaped bluish bacteria in the haematoxylin and eosin sections
2. Presence of empty spaces (gas bubbles) of various sizes in the affected tissues
3. Changes as seen in coagulation necrosis or liquefaction necrosis can be found in the dead gangrenous tissues.

Gas gangrene

It is produced by several species of the genus *Clostridium.* These organisms grow in dead or living tissues of the body. Organisms growing in the dead tissues are saprophytes or putrefactive forms. *Cl. chauvoei* causes black quarter (BQ) in cattle. This organism kills the tissues, and grows in the dead tissue to produce gas and edematous changes. Other organisms like *Cl. welchii, Cl. septique, Cl. oedematiens* produce also gas gangrene following their intramuscular administration in animals. Malignant odema is also an example of gas gangrene.

Effects

1. Absorption of highly toxic decomposition products into patient's body has disastrous effects which may cause death.
2. Sapraemia may occur. Putrefactive bacteria enter the blood and may be disseminated throughout the body in different organs.

Hyaline (Hylos, Greek = glass or glass like) degeneration

Hyaline degeneration is a condition in which the tissues are transformed into clear, homogenous or glass like substance. The hyaline may be smooth, solid or dense in consistency.

There are two kinds of hyalines, namely (i) connective tissue hyaline and (ii) epithelial tissue hyaline.

i. Connective tissue hyaline

This hyaline substance is a protein in nature and resists chemicals and solvents. Connective tissue fibres of the smaller arteries or capillaries are affected by hyaline degeneration. Old blood clots, renal casts, tumour and scar etc. are well known to undergo hyalinization. Spleen, renal glomeruli, central nervous system, thyroid and lymphatic glands are most frequently affected by hyaline degeneration.

Staining reaction

Acid stains (oxyphile) stain hyaline material. Van gieson's picro-fuchsin stains old hyaline yellow and fresh hyaline red.

Causes

Hyaline degeneration is associated with conditions like tuberculosis and deficient nutrition etc. Glomeruli in inflammatory condition like nephritis and lymph glands draining tuberculous lesions show hyaline degeneration. Deficient nutrition is another factor for causing hyaline degeneration. This is noticed in the walls of blood vessels of uterus and ovary in old age. Glomerulus shows hyalinization due to interference with blood supply. Renal casts and thrombi are affected by hyalinization and uterine fibroma also shows hyalinization.

Gross appearance

It is a white, shiny, glossy, solid or dense substance. Hyaline is a smooth homogenous and structureless to the naked eye. It is acidophilic and stains brilliant pink with the usual stains.

Causes

1. Lack of nutrition. There is poor blood supply in large scars
2. Toxins or injurious substances.

Microscopical appearance

Connective tissue elements in the smaller blood vessals are affected by hyaline degeneration but the changes are most often marked into inner coats of the smaller arterioles. The fibers swell up, loose their normal structure and look like homogenous substance. Capillary also shows similar changes. Other cellular tissues also look homogenous glass like substances and stain with acid stains. There are no nuclei and fibrils in completely hyalinised tissues and one sees gradual transition of the hyalinised tissues into surrounding fibrous tissue.

2. Epithelial hyaline

The examples of pathological hyalines are corpora amylacea (Latin for starchlike bodies) found in the prostate and lungs. Coropora amylacea are round, opaque bodies found in the damaged areas of the white matter of brain.

Gross appearance

Hyalines are not detectable grossly.

Microscopically these are found as homogenous and concentreically laminated bodies which stain pink.

Keratohyaline

Keratin is also a hyaline and its transformation is called keratinisation or cornification as noticed in epidermal cells. Keratohyaline may be normal or pathological. Pathological keratoyaline is found in excess at abnormal location. Cornified layer of stratum corneum in epidermis is a normal feature.

Gross appearancc

It is hard, colourless more or less transparent horny material. In the skin, it looks wrinkled.

Microscopic appearance

It is solid and homogeneous and looks like a pink staining mass in the skin. It blends gradually with the prickle cell layer. Examples of abnormal keratinisation are callus, corns, warts and epithelial pearls in squamus cell carcinomas.

Hyaline degeneration is followed by the following changes :

1. **Calcification**

2. Hyalinised tissue has less strength and functions less properly than the normal tissue.

It is inelastic and inflexible. A permanent irreversible degenerative change in the tissue is its important feature.

Glycogenic infiltration

Glycogenic infiltration is a condition in which glycogen appears in abnormal amount in cells which normally do not appear to contain it. In the normal animal, it is noticed in the liver cells as particulate glycogen. The muscles also contain it.

In diabetes mellitus, glycogenic infiltration occurs in the cells of Henle's loops and heart also contains excess of glycogen. The deposition of glycogen in the cardiac muscle is associated with high concentration of glucose in the blood. In glycogen storage disease characterized by a derangement of carbohydrate metabolism, there is an excessive deposition of glycogen in the kidneys, liver and heart muscle cells.

Pathologically, it occurs in the cytoplasm of epithelial cells of the liver, kidneys, leucocytes, cardiac muscle, smooth muscle cell, spleen, lymph nodes and brain.

Gross appearance

Glycogen is not detected grossly,

Microscopic appearance

Microscopically, glycogen is seen in the cytoplasm of cells as clear spaces or vacuoles. These spaces are areas of glycogen deposits which have been dissolved away due to water used in the ordinary method of washing. In well nourished animals, hepatic epithelium has considerable glycogen which makes the cytoplasm foamy in appearance.

Staining

The tissues are fixed in absolute alcohol and alcoholic stains are used through out the processing and staining of the tissues. Glycogen stains bright pink or red with Best's carmine.

Significance

Glycogen in the tissues is indicative of some other diseases.

Amyloid infiltration (Lardosis)

Amyloidosis refers to an infiltration of a firm and solid extacellular substance called amyloid in various organs. It is a protein that contains somewhat less than 5% carbohydrate. An inert waxy substance appears in artery, veins and into the middle coat of the smaller arterioles and supendotheial layers of the capillaries. It seems that an amyloid is related to some derangements or abnormalities in the immune system and is considered by some pathologists as a disease of immune system. It affects a single organ or even a system. Amyloid is a proteinacious substance which is deposited between cells of organ. Light microscopy reveals a hyline, eosinophilic, amorphous, extracellular material. An extracellular excessive deposition of amyloid produces pressure atrophy in the affected adjacent parenchymatous cells in organs. Congo red stains amyloid as a red and pink substance. Polarizing microscopy reveals a green birefringence of the stained material. It is noticed in several diseases and similar appearing proteins are present between the cells of affected organs.

The physical nature of amyloid is as follows

i. Non- branching fibrils of indefinite length and diameter of 7.5 to 10 nm by electron microscopy

ii. Cross β-pleated conformation by X-ray crystallography

iii. Green Birefrigence of the β-pleated sheet by polarizing microscopy

Human patients of myloma reveal Bence-Jones protein (AL amyloid) in serum or urine.

Some other proteins found in amyloid are

i. Mutant forms of transthyrein C (a serum protein)

ii. β_2- microglobulin (a serum component of MHC class molecules

iii. β_1-amyloid protein (Aβ)

iv. Hormone precursors like procalcitonin and proinsulin

The man types of amyloidosis (Table 3) are as follows :

Table 3- Classification of amyloidosis

Type	Associated Disease	Major fibril protein	Releated precursor protein
Primary amyloidosis (immunocyte dyscarasias with amyloidosis). Reported in man dog and cat	Multiple Myeloma, Plasmacytoma and β- cell lymphoma	95% of amyloid material. Amyloid light chain (AL) Proteins produced by immu-no globulin secreting cells	Immuno globulin light cha-in (λ type)
Secondary amyloidosis Noticed in man dogs and cats	Chronic inflammatory diseases like TB and glanders etc.	Amyloid associated material (AA) derived from precursors in the SAA protein	SAA in (serum amyloid associated protein)
Localised amyloiidosis reported in man and dogs	Senescence	Transthyretin	

Occurrence

This disorder can affect any organ. The spleen, liver, kidneys, lymphnodes, etc. are noteworthy sites. It is a slow progressive condition and can replace the affected tissues. It has been found associated with prolonged suppurative conditions affecting bones or other tissues and also with chronic bacterial diseases (e.g. tuberculosis and glanders) and experimental conditions as seen in horses for production of antitoxins. Rabbits and chickens following injections of pyogenic organisms or turpentine show amyloid infiltration in the tissues.

Gross appearance

Small amount of amyloid cannot be visible to the naked eye. Larger accumulations of amyloid are firm, white, opaque and look like lard. Organs other than the spleen show slight change in their appearance. The liver becomes larger, paler and firmer than normal. The edges are rounded and the cut surfaces have clear and waxy appearance. In amyloidosis, spleen shows two kinds of lesions, namely (i) sago spleen and (ii) bacon spleen.

1. Sago speen

There can be some enlargement of the spleen which shows Malpighian bodies as projecting boiled sago grains.

2. Bacon spleen

In this condition, the amyloid appears in the splenic pulp and when the spleen is cut or sliced, the cut surfaces look like bacon.

Microscopic appearance

Amyloid is a dense and nearly homogenous material and it replaces and obliterates pre-existing cells. The borders of amyloid are sharp which distinguishes it from connective tissue hyaline. It stains purplish pink in haemotoxylin and eosin sections. Congo red stains amyloid orange. Thioflavin T staining imparts fluorescence to amyloid with greater specificity than congo red. Cresyl violet stains amyloid violet whereas methyl violet stains it pink against a violet background. On fresh slices of tissues, application of iodine brings out the amyloid as a clear brown material. In liver, amyloid infiltration is seen in the spaces of Disse (fig.8). Amyloid accumulates in the glomeruli in renal amyloidosis. Amyloid infiltration is noticed in the vicinity of the smaller blood vessels in most of the organs like spleen, liver and kidneys etc. Splenic follicles and the walls of sinuses and connective tissue framework in the red pulp are infiltrated with amyloid. Heart reveals focal subendocardial deposits of amyloid. Such deposits are also noticed between the cardicle muscle fibres.

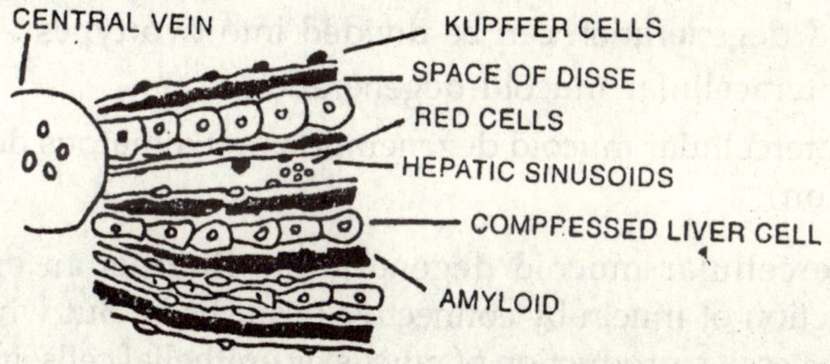

FIG. NO.- 8 AMYLOID INFILTRATION

Causes

1. Occurrence of secondary amyloidosis in association with many chronic diseases.
2. Production of amyloid by reciculoendothelial cells, histiocytes and plasma cells etc. in some pathological states.
3. Origin as a degeneration product of immunoglobulin.
4. Primary amyloidosis is independent of any cause.

Results

1. Fatty changes and atrophy in the hepatocytes due to pressure and narrowing of the blood vessels due to amyloid deposits.
2. Nephritis following kidney lesions.
3. Rupture of liver and death in the horses. This has been noticed in horses used for antitoxin preparation.

Mucoid degeneration

Mucoid degeneration is a condition in which mucus (mucin) is produced in excess by certain kinds of tissues. Mucous membranes and certain connective tissues normally produce mucus or mucin which is a clear, viscid, slimy substance. Mucus is produced by the lining cells of the respiratory and gastrointestional tract and also from portions of the genital tract. It is normally found in the umbilical cord, synovial cavities and bursa of Fabricius.

Mucoid degeneration can be divided into two types :-

I. Interacellular mucoid degeneration.

II. Intercellular mucoid degeneration (myxomatous degeneration).

In intercellular mucoid degeneration, there is an excessive production of mucin by connective tissue cells. But when there is an excessive production of mucus by epithelial cells, it is called intracellular mucoid degeneration.

Intracellular mucoid degeneration :

It is noticed in the inflammations of the mucous membranes. Excessive mucus is produced in common cold, tracheitis, bronchitis, gastritis, entritis and certain kinds of cancer of the gastrointestinal tract and breast. In mucoid cancer, mucus is collected in cysts. The mucous cells lining the cysts disintergrate and get lost in the homogeneous contents. Mucus stains blue with haematoxylin.

Gross appearance

Excessive amount of mucus as clear or slimy material is found on the mucous membrane or in tissues.

Microscopic appearance

Bluish or grayish strands of mucin can be seen clinging to the mucous membranes in haemotoxylin and eosin sections. Increased number of goblet cells and hyperaemia etc. are seen in the mucous membrane.

Intercellular mucoid degeneration (Serous atrophy of fat)

It occurs in tissues containing some fibrous or adipose tissue elements and has been noticed in skeletal muscle, bone marrow or around the coronary groove of heart in many wasting conditions or state of inanition.

Causes

1. Blood borne toxins and chronic wasting diseases like tuberculosis and parasitic infections.
2. Poor nutrition.

Gross appearance

1. The affected tissue looks like translucent and watery or shiny material in covering of serosae etc.
2. It also resembles Wharton's jelly.

Microscopic appearance

There is a proliferation of connective tissue of embryonic characteristics and its intercellular material has bluish tinge and scanty fibrils with crisscross arrangement and the connective tissue cells have hyperchromatic nuclei and are stellate (i.e. star sharped or triangular) in appearance.

Significance

1. Presence of some inflammatory conditions of the mucous membranes.
2. Poor nutrition or infection of the animals with some wasting or cachectic diseases. (e.g. TB, cancer and stomach worm infection)
3. It indicates the state of inanition.

PATHOLOGIC PIGMENTATION

It refers to an abnormal deposition of pigments which are substances having colour of their own in the tissues and are detectable by the naked eye or with aid of microscope. These pigments can be divided into the following types:

(1) Exogenous and (2) Endogenous pigments.

(1) Exogenous pigments (i.e. the pigments formed outside the body)

(a) Carbon

(b) Dusts

(c) Metals

(d) Tattoos

(e) Kaolin

(f) Carotenoids

(2) Endogenous pigments (i.e. the pigments formed inside the body)

A. Phenolic pigments
   1. Melanin

B. Autogenous pigments
   1. Haemoglobins
   2. Haematins
   3. Haemosiderins
   4. Bile pigments

C. Lypogenic pigments
   1. Tissue lipofuscins
   2. Ceroid

Anthracosis (Carbon particles in the tissues)

This refers to presence of carbon particles as black pigments in the tissues. Carbon can be found in the lungs and lymph nodes and is found in man and animals working or living in smoky cities and coal mines.

Gross appearance

Moderate or large amounts of coal particles impart mottling or speckling with black or grey discolouration to the lungs. Lymph nodes are darkned.

Microscopic appearance

Carbon appears as minute black granules seen either among cells or in their cytoplasm. In the lungs, it is seen in the cells, macrophages, alveolar walls and connective tissue septa. In the lymph nodes, it is seen between lymphoid cells and inside large mononuclear cells.

Results

Reasonable amounts of carbon do little harm and cause no symptoms. Excessive amounts cause fibrosis in the lungs or predispose lungs to many kinds of infections (e.g. TB).

Dusts, silicon-dioxide, iron dust, asbestos, silver and lead and tattoos

When dust is inhaled and retained in the lungs, the condition produced is called pneumoconiosis, Silicon dioxide is inhaled in rock quarries and mines. These pigments occur as fine crystals and cause proliferation of connective tissue elements in the lungs. Grossly, the lungs become firm and show nodules. Iron dust is inhaled as iron oxide in mines. Microscopically, these occur as red crystals of varying sizes in the lungs and can be demonstrated with the Prussian blue reaction. Presence of iron in tissue is called siderosis. The affected tissues stain blue when such tissues dipped in Potassium ferrocyanide are treated with weak HCl.

Asbestosis refers to presence of asbestos in the tissues of body. Microscopically, it occurs as fine fibres or bodies. Asbestos causes fibrous scarring of pleura around bronchioles and within alveolar septa. Foreign body giant cells are found adjacent to the asbestos particles.

Silver (causing argyria) can be deposited in the tissues. Use of argyrols to treat infection in humans give rise to a condition called argyria and humans develop a peculiar ghastly grey colour. Micrscopically, it is deposited in the connective tissue extracellularly in the upper corium. Microphages may also reveal silver.

Lead poisoning occurs due to prolonged ingestion or assimilation of small amounts of lead compounds. A blue black discoloration is imparted to gums to form what is called a **lead line** as a characteristic sign of this poisoning.

Tatoos

Various kinds of pigments are used to produce tattoos. These pigments include India or China ink etc.

These pigments are inert and show no tissue reaction and are seen extracelluarly between connective tissue cells.

ENDOGENOUS PIGMNTS

Haemotagenous pigments

A variety of pigments are derivatives of haemoglobin. Haemosiderin and bilirubin are normal physiologic pigments whereas others are pathologic like methaemoglobin and haematin etc. formed by protozoan and metazoan parasites.

Haemosiderin

It is a shiny, golden yellow and golden brown pigment derived from haemoglobin. It is found at the sites of old haemorrhages, receding corpora haemorrhagica and chronically congested lungs.

Microscopically, it occurs as golden coloured spherules, 2 or 3 microns in width or diameter engulfed by phagocytes. The presence of haemosiderin in tissues is determined by the Prussiana blue reaction. The affected organs assume brownish colour following deposition of iron pigments.

Haemosiderosis arises from haemolysis. Haemoglobin is seen in the phagocytes of spleen in haemolytic anaemia. Kupffer cells in the livers show haemosiderin. In chronic passive congestion of the lungs, haemosiderin is found in macrophages in the alveolar spaces of the lungs. These macrophages are called heart failure cells. The affected lungs become leathery, some what hard and brown in colour due to haemosiderin in the alveoli and this condition is known as brown induration of the lungs arising from some valvular diseases (e.g. stenosis or heart failure).

Haemochromatosis

It is a rare condition in human beings in which a pigment indistinguishable from haemosiderin is found in the cytoplasm of the epithelial cells of the liver, pancreas, kidneys, spleen and various other organs. In goats, heavy deposition of haemosiderin like pigment has been reported in kidneys, Kupffer cells of liver and reticuloendothelial cells of the spleen and lymph nodes. Such goats are not clinically ill.

Methaemoglobin

It is an oxide of haemoglobin (ferric iron) which is dark red. It is produced by poisonings with nitrites, chlorates and some other compounds.

Carboxyhaemoglobin

It is a bright cherry red substance which is a combination of carbon monoxide with haemoglobin.

Haematins

These are pigments formed by the action of acids on haemoglobin. Hydrochloric acid forms haematin which is often seen adjacent to gastric ulcers following haemorrhages.

In malaria, haematins are seen in the parasitized cells and in the cytoplasm of reticuloendothelial cells in the spleen, lymph nodes, liver and bone marrow.

Bile pigments

Red cells are undergoing normal destruction in the body. The iron and globin are reused by the body and the porphyrin is changed into a soluble pigment bilirubin by reticuloendothelial cells in the bone narrow and spleen, etc. Bilirubin circulating in the blood reaches liver where it is excreated in the bile. When there is an excessive destruction of red cells, it accumulates in body and gives rise to the condition of jaundice. There is staining of body fluids and tissues with bile pigments which are iron free substances.

Jaundice (Icterus)

It denotes the staining of certain body fluids and tissues integuments with bile pigments present in the circulation (cholemia). The bilirubin arising from conversion of porphyrin into billirubin and its oxidation product biliverdin are called bile pigments. The reticuloendothelial cells in the liver, bone marrow and spleen cause breakdown of the haemoglobin and give rise to formation of bile pigments. Biliverdin is a green pigment where as bilirubin is an orange yellow pigment. The liver converts haemobilirubin into cholebilirubin and conjugated bilirubin (bilirubin diglucuronide arising from its conjugation with glucuronic acid). Bile pigments are converted into urobilinogen (a reduced colourless product) and stercobilin (a brown pigment) of faeces by intestinal bacteria.

Bilirubin and urobilinogen are reabsorbed in the intestine, and the bile pigments and urobilinogen enter into blood stream and finally return to liver parenchyma. The Van Den Bergh's test differentiates cholebilirubin (i.e.bile pigment formed by passing through liver cells). When a solution of sulphanilic acid and sodium nitrite is added to serum containing bilirubin diglucuronide (conjugated bilirubin), cholebilirubin or hepatobilirubin, a bluish violate colour develops at once. This is called the immediate or direct reaction. If there is a presence of haemobilirubin, the reaction is delayed but the addition of alcohol to serum before mixing with reagents causes development of an immediate bluish violet colour. This type of reaction is called indirect or delayed reaction. The direct reaction is seen in the obstructive jaundice due to influence of the liver action but indirect reaction occurs in the haematogenous jaundice. Haemobilirubin and cholebilirubin are found in the serum in toxic jaundice and there is, thus, development of a biphasic reaction.

Three important sites (Fig. 9) of icteric causes noticed in the body are as follows :

1. Red cells in the blood stream

Piroplasms and *Clostridium haemolyticum* destroy the red cells and cause icterus due to formation of haemobilirubin.

2. Hepatocytes in the liver

Toxins, infections and poisons destroy or damaged the liver cells to cause jaundice.

3. Extrahepatic bile duct

Stones, flukes and worms obstruct the bile duct and cause icterus due to a bile pigment called cholebillirubin.

4. Intestinal bacteria convert bile pigments into stercobilin (a brown pigment of faeces) and urobilinogen (a reduced colourless product). Bilirubin and urobilinogen are reabsorbed into the blood stream in the intestine and reach the liver

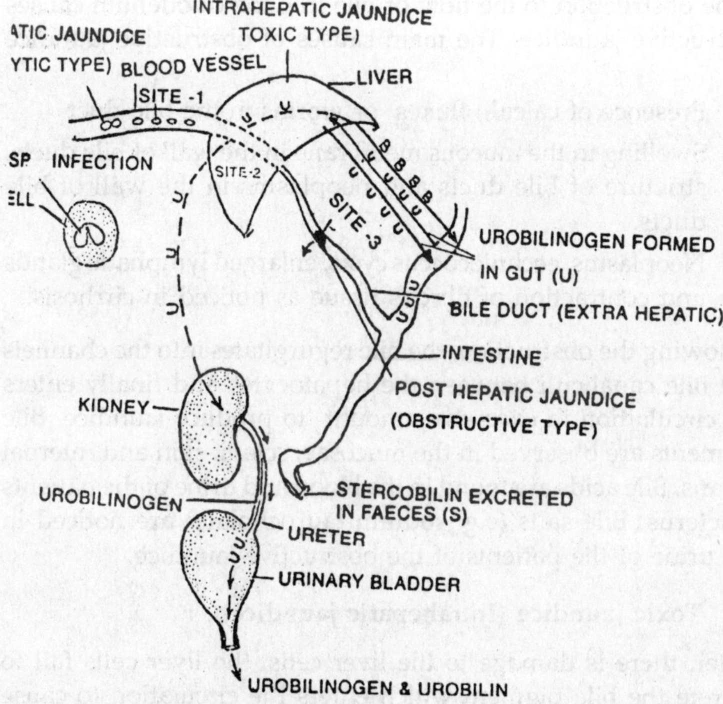

Fig. 9. Sites of icteric causes and formation of urobilinogen from bilirubin in the intestine.

The types of jaundice based on its pathogenesis are as follows :

(1) Hepatogenous jaundice

The conjugated bilirubin arises from conjugation of bilirubin with glucuronic acid in hepatocytes to be excerted in bile as bilirubin diglucuronide.

(2) Haematogenous or haemolytic jaundice (prehepatic jaundice).

Hepatogenous jaundice is further subdivided into (a) an obstructive jaundice and (b) toxic jaundice.

(a) Obstructive jaundice (Post hepatic jaundice)

Some obstruction to the flow of bile into the duodenum causes obstructive jaundice. The main causes of obstructive jaundice are :

(i) Presence of calculi, flukes or worms in the bile duct

(ii) Swelling in the mucous membrane in the wall of bile ducts, stricture of bile ducts and neoplasms in the wall of bile ducts.

(iii) Neoplasms, ecchniococcus cysts, enlarged lymphatic glands and contraction of fibrous tissue as noticed in cirrhosis.

Following the obstruction, the bile regurgitates into the channels like bile canaliculi between the hepatocytes and finally enters the circulaltion in excessive amount to produce jaundice. Bile pigments are observed in the mucosae, sclera, skin and internal organs. Bile acids are found in the blood and urine of the patients of icterus. Bile salts (e.g. sodium taurocholate) are noticed in the urine of the patients of the obstructive jaundice.

(b) Toxic jaundice (Intrahepatic jaundice)

When there is damage to the liver cells, the liver cells fail to excrete the bile pigment which enters the circulation to cause icterus marked by deposition of bile pigments in the Malpighian layer of skin and mucosa etc. Infective and septicaemic conditions cause toxic jaundice. Leptospirosis produces toxic jaundice due to damages caused by the leptospira in the liver. Both kinds of bilirubin (prehepatic haemobilirubin and post hepatic cholebilirubin) are found in the blood of patients of toxic jaundice.

In toxic jaundice, both haemobilirubin and cholebilirubin are found in excess in the patient's blood. Presence of cholebilirubin is also noted in the urine of the patients affected with toxic (intrahepatic) jaundice.

Haematogenous jaundice

It arises from excessive haemolysis of the red cells as seen in piroplasmosis, streptococcal and *Clostridium haemolyticum* infection. The damaged liver is unable to cope with increased break down of red cells (haemolysis) and excessive amount of haemobilirubin accumulates in the blood to cause the symptom of icterus. Bile acids are found in the blood and urine of the patients of icterus. Bile salts, (e.g. sdium taurocholate) are noticed in the urine of patients of the obstructive jaundice.

In short, icterus is a symptom but not a disease arising from different or diverse kinds of aetiological factors like bacteria, parasites and viruses etc. producing diseases in the animals.

Diagnosis

It is based on symptoms and presence of bile pigments in the blood and urine. Van den Bergh's test is done to find out the kinds of jaundice in the patients. The kinds of reactions in different types of jaundice are given below :

1.	Obstructive jaundice (i.e. (extrahepatic form)	Direct reaction due to post-hepatic bile pigment arising from some liver action. Both the cholebilirubin and the conjugated bilirubin (bilurubin diglucucronide) give a direct reaction.
2.	Haemolytic	Indirect reaction due to the presence of jaundice pre-hepatic bile pigment (not subjected to some liver action)
3.	Toxic jaundice	Both direct and indirect reactions due to presence of pre-hepatic and post-hepatic bile pigments

Tests are also done to find out icterus index and bilirubin (urbilin) in the urine (choluria). In haemolytic jaundice, large amounts of urobilibogen are found in the faeces and urine. But there can be almost no urobilinogen in the faeces and urine in the patients of obstructive jaundice.

An yellowish substance called haematoidin is found at the sites of old haemorrhages. This is belived to be the same as bilirubin. Bilirubin and haematoidin are negative to the usual test for iron. Haematoidin appears as angular yellow crystals and are seldom phagocytised in the tissue.

Effects

Itching, headache, fatigue and excitability are symptoms noticed in humans. In experimental jaundice (e.g. dogs), one sees jaundice and intoxication in dogs on the 4th or 5th day after ligation of the bile duct. Irritability, diminished blood clotting, itching and depression are other signs noted in such experimentally induced canine jaundice.

Melanin

It is an endogenous pigment and autogenous in nature and gives black colour to the skin and hair etc. and is formed by special cells called melanoblasts. It occurs in skin, iris, dendritic cells of the piaarachnoid and substantia nigra of the brain. It is a very common pigment which is small and brown in colour and free from iron. Melanosis is a deposition of melanin in various organs like lungs and aorta etc. of an individual. Internal organs of the herbivores and mammary glands of black gilts show areas of melanosis. Melanosis is also found in the tumours like melanoma and may appear in the urine (i.e. melanuria).

Lipogenic pigments

These tissue pigments are coloured substances derived mainly or partly from lipids. Two pigments, namely, lipofuscin and ceroid are of pathological significance. Lipofuscin (lipo, Greek for fat and fuscus, latin from dusky and dark).

A pigment (lipochrome in nature) is also called wear and tear pigment i.e. the pigment of the brown atrophy of heart. It has been noticed in heart, skeletal muscle, liver and renal glands, neuroma, thyroid, parathyroid, kidney, ovary and testicle etc. It is Sudanphilic (i.e. stains with fat stains) and acid-fast and negative to iron stain and is found in cachectic or senile atrophy e.g. a brown atrophy of heart.

Gross appearance

Lipofuscin imparts a brownish tinge to cardiac and skeletal muscles.

Microscopic appearance

The pigment appears as minute yellowish or dark brown granules. In heart muscle, it is noticed chiefly near the poles of the nuclei and in skeletal muscle, it is found in any part of the fibre. In liver and adrenal glands, it is distributed throughout the cytoplasm.

Significance

Its presence in cells of organs indicates some wasting diseases, senility and emaciation etc.

Ceroid

It is a pigment formed from unsaturated lipids and related to choline deficiency. Macrophages and hepatocytes reveal their presence.

A pigment which is essentially identical to ceroid occurs in vitamin E deficiency. It is found within the macrophages throughout the body, fat cells, cardiac muscle and smooth muscle cells of the spleen and intestine. Microscopically, it is granular or homogenous and yellow or brown in colour. The pigment is resistant to fat solvents and Sudanophilic even after paraffin embedding procedures. This pigment is also acid fast and negative to fat stains. Ceroid is indistinguishable histochemically from lipofuscin.

PATHOLOGIC CALCIFICATION (CALCAREOUS INFILTRATION)

Calcification in tissues or organs may be either physiological or pathological in nature. Pathological calcification refers to the deposition of calcium salts at the sites other than the normal ones. Calcium in body is obtained from blood and execreted in urine and large intestine and its level in health is about 10mg per 100cc. of blood. Parathyroid hormone maintains level of calcium in the blood by withdrawing it from the skeleton. Hypocalcaemia is found in case of hyperactivity of parathyroid. There is an increased excitability of motar nerves (twitching and convulsions) in hypocalcaemia. Vitamin D regulates the exchange of calcium between blood and skeleton.

Causes

1. Alkalinity augments calcium deposition. Carbon dioxide production is low or not seen in necrosis, degeneration, chronic inflammation, scars and infarcts. This causes the medium to be alkaline and deposition of the calcium and phosphate ions are favored.
2. Increase in the level of alkaline phosphatase. Calcium phosphatase is increased abnormally in certain types of injured tissues. This leads to an increased production of calcium and phosphate ions locally beyond their solubility with resultant calcification.
3. Formation of fatty acids and glycerol. Fat splits into fatty acids and glycerol. Fatty acids combine with calcium to form calcium soaps in which soaps are, later replaced by carbonic and phosphoric acids to form calcium salts.

Pathologic calcification is divided into dystrophic calcification, metastatic calcification and calcinosis.

1. Dystrophic (Gr. = ill + to nourish)

Calcification occurs in dead, dying and degenerating tissues without regard to the concentration of calcium in the blood stream. It is the commonest type in old tuberculous tissue, intima in the hardening of the arteries (arteriosclerosis), media of large arteries, cartilage of ribs and the trachea in old age. Inspissated pus and empyema, old thrombi, degenerating tumours, castration wound, parasites and foreign bodies etc. are the common sites of calcification constituting examples of dystrophic calcification.

2. Metastatic calcification

It occurs due to hypercalcaemia owing to some diseases like hypervitaminosis D in aorta,coronary vessels and heart muscles of rat. It is also found in the cases of renal failure and hyperparathyroidism.

3. Calcinosis

It is an abnormal deposition of calcium salts in various parts of the body. Widespread calcification of tissues has been seen in several species of cattle, sheep and goats etc. Enzootic calcinosis has been reported in Germany and India.

Staining reaction

Calcareous material in tissues stains deep blue by haematoxylin and black by 5% silver nitrate solution.

Gross appearance

Deposits of calcium look like white granules, clumps or large plaques of extremely firm, gritty and uncutable material.

Microscopic appearance

The calcium carbonate or phosphate is found as irregular granules of micrscopic size or little larger. With basic haematoxylin, it stains purple and calcium salts can be demomstrated in tissues by Von-kossa's method.

GOUT OR URATIC INFILTRATION

It is a condition in which crystals of uric acid or urates (e.g. sodium biurate) are deposited in the tissues.

Occurrence

It occurs in man and birds. Sharp crystals appear in the articular and periarticular tissues. These crystals act like irritating foreign bodies causing intense pain and acute inflammation. In this, there are fluctuations in protein or purine metabolism. In birds, two forms, namely, visceral and articular forms of gout are found. Absence or low level or uricase is related to occurrence of gout. Apes and human beings do not possess it but most animals have hepatic enzyme called uricase which metabolises relatively insoluble uric acid to the more soluble limits before its exertion in urine. Impaired renal excretion of uric acid seems to be associated with gout in birds and acute inflammatory reaction is produced in the joints.

Gross appearance

White chalky masses are found in the tissues in gouty joints in chickens and turkeys. These are called tophi which may occur in the subcutaneous tissues and ulcers are formed. In visceral gout, the serous surfaces of the body cavities i.e. pericardial sac are encrusted by a thin grayish layer of urates having a metallic sheen.

Microscopic appearance

White chalky masses are found in the tissues in gouty joints in chickens and turkeys. These are tophi which may occur in the subcutaneous tissues and ulcers are formed. In visceral gout, the serous surfaces of the body cavity i.e. epicardial sac are encrusted by a thin grayish layer of urates having a metallic sheen.

Microscopic appearance

Inflammatory infiltrations including presence of macrophages and foreign body giant cells in conjunction with clusters of sharp acicular crystals are located in an particular surface (joint capsule). In turkey, deposits of urate crystals are found in the kidneys. The urates are surrounded by macrophages and multinucleated giant cells in the renal tubules (Fig.-10)

Fig.10 Renal gout. Note the deposits of urates surrounded by macrophages and foreign body giant cells H & E x 400

Significance

There is a recovery with change in the diet given to birds.

PATHOLOGIC OSSIFICATION

It refers to the formation of a true bone at other than normal sites. The bone is formed by osteoblasts and histologically, there is a formation of adult bone with lamellae, lacunae and even bone marrow. Ossification of lateral cartilage of horse's foot, the tracheal or laryngeal cartilage and bony spicules in alveolar septa are some of its examples. Prolonged irritation may cause pathological ossification.

Microscopical appearance

Cells, lacunae and calcified acellular areas are found.

Results

1. Interference with movement of limbs and organs etc.
2. Interference with exchange of oxygen in the lungs.

CONCRETIONS OR CALCULI

The term lithiasis means formation of concretions or calculi. In this, there is a precipitation of calcium salts in certain secretions like urine or bile etc. The presence of a foreign body serves as nucleus or nidus for calculi formations. They are found in the different organs of the body like kidney, gall bladder, urinary baldder, and intestines etc. Some of the main kinds of calculi or stone formations are as follows :

1. Uroliths (Urinary calculi)

These are stones formed anywhere in the urinary passage of the body (i.e. from pelvis to urethra) and the process of its formation is called urolithiasis. Cystic calculi are stones formed in the urinary bladder. They can also be found in ducts of Bellini (terminal urinary tubules), the renal pelvis or ureter. Stones lodged in the ureter or renal pelvis produce disuse or pressure atrophy of the affected kidney or can lead to the state of hydronephrosis. Uroliths may look like sand particles or may be big enough to fill the bladder or renal pelvis. Stones may be white or yellow, smooth or rough or may be even faceted or rounded. The faceted calculi have flattened sides. Stones may measure a few millimeters in diameter or may be big enough to obstruct the sigmoid flexure of urethra in cattle. Uric acid calculi consist of ammonium and sodium urates and uric acid. These may be yellow or brown, firm or hard in consistency. Phosphate calculi are white or grey and have chalky consistency. They may be soft or friable and can be crushed with fingers. Phosphate calculi are found in herbivores and calculi of magnesium ammonium phosphate are commonly found in dogs.

2. Intestinal calculi (Enteroliths)

Intestinal calculi are found in horses at sites like large colon and caecum rarely. The calculi may be single or in pyramidal form with facets in several small forms. They may be spherical, ovoid, cylindrical or pyramidal in shape. The calculi may reveal nuclei in sections around which substances can be found deposited in concentrically arranged laminae. They consist of ammonia, magnesium phosphate and calcium salts, epithelial cells and other organic matter. False calculi do not possess minerals but vegetable fibres, hair etc. bound by mucus. Bran or food and water rich in mineral salts are predisposing factors for these stones.

Causes of calculi

These are as under :

1. Excessive amounts of salts in the urine
2. Retention of urine
3. Reduced intake of fluid and excessive loss of fluid
4. Ammoniacal fermentation in urinary bladder
5. Increased acidity or increased alkalinity
6. Presence of a foreign body
7. Certain foods like bran and water in stone districts predispose to urolithiasis.
8. Calcium phosphorus imbalance
9. Deficiency of vitamin- A in feeds

Some types of calculi are as follows.

1. Renal calculi are small, numerous sand like partices or coarse grit.
2. Pelvic calculi may be single large calculus.
3. Cystic or vesical calculi may be small and numerous. Calculi may be rough, smooth, nodular, hard, brittle, laminated and white or dark brown in colour.The compositions of the calculi are given below.

Animals	Nature of calculi
Herbivores	Silicates, carbonates and phosphates of calcium and magnesium.
Carnivores	White or yellow phosphates, urates and oxalates. Oxalate stone is very hard and heavy and has sharp edges.
Swine	Phosphhates, oxalates and carbonates.
Sheep	Xanthine calculi, fragile laminated or brownish red in colour.
Dog and cat	Cystine calculi are small. Soft with greasy shiny appearance. Yellow in colour.

Effects

1. Stones cause inflammation of the organs or at the sites of their lodgment .
2. Urine retention, uraemia, hydronephrosis and hyperthophy may be noticed in the patients.
3. Obstruction may be caused.
4. Formation of sac or pouch by large calculi which are often less dangerous.
5. Impaction by the small calculi is seen in the narrower portions of the intestine and obstruction caused is invariably fatal.
6. Ulceration in the intestinal mucosa in the vicinity of the calculi.

3. Billiary calculi or Gall stones

These are reported mostly in cattle but rarely in dogs. The term cholelithiasis refers to the formation of gall stones in bile in the billiary passages. They are more frequently found in the gall bladder. They have been found in bovines, chickens and pet animals.

Causes

1. Concentration of bile.
2. Infection in the biliary passages.
3. Bile stagnation. This leads to the crystallization or deposition of some of its constituents.
4. Precipitation of calcium is favoured by parasites etc. and the salts get deposited around desqamated cells acting like nuclei.

Forms

1. Single or multiple in various sizes. These may look like peas or oranges.
2. The may be oval, spherical or irregular in forms with rough or smooth surfaces.
3. Often faceted or pyramidal stones are formed due to fric-

tion amongst several such stones.

4. Yellowish brown, red or white in colour and the cut surface has laminated appearance like that of an onion.

Composition

1. Cholesterol is a common constituent in human beings but it exists in small proportion in cattle.
2. Bile pigments, fibrin or calcium may be present.

Effects

1. Cholecystitis and cholangitis and even cancerous changes may follow.
2. Bile statis and jaundice in case of impaction in the ducts.
3. Stones may have silent existence or produce nausea, vomiting and very painful state as reported in humans.
4. Pancreatic calculi (Pancreoliths).

These stones are found in the pancreatic ducts of cattle. They are hard, white or numerous and may be small in size, and consist of carbonates and may be small in size, and consist of carbonates and phosphates of calcium and magnesium in association with organic materials. These stones produce inflammatory state in the pancreatic ducts.

5. Sialoliths (Salivary calculi)

Sialoliths are stones found in ducts or in the salivary glands. A chronic inflammatory state of the salivary gland may lead to formation of calculi. Calcium salts are precipitated on the desquamated cells or condolidated exudates in the ducts of glands and further addition of salts leads to increase in size of the calculi. Foreign bodies may lead to deposition of salts over it. An obstruction in the duct due to calculi may lead to atrophy of the gland. A dilated duct may give rise to a cystic state in the gland. Ranula refers to a cyst formed in the sublingual duct located in the frenum linguae.

6. Phytoconcretions

These are concretions made up of plant fibers. The compact or solidified unorganized material in them will be either endogenous or exogenous in origin and is located in hollow organs. Hair with other materials like mucus, mineral salts etc. forms what is called piliconcretion. Hairs in the stomach or intestine get rolled in the form of balls which are encased in smooth and glistening shell like structure of concretions (hair halls). The concretions containing plant fibres infiltrated with triple phosphates are called phytoconcretions (food balls). Their outer surface is velvety and such balls are spongy and soaked with fluid. Fissures or cavities containing decomposing food material are noticed on the cut surfaces of such ball concretions. These have been reported in horses. Piliconcretions are found in cattle and pig.

7. Caproliths

These are faecal concretions noticed in dogs and birds and are found in cases of chronic constipation and actually represent hard masses of inspissated faeces in parts of the digestive tract like small intestine, colon and rectum. Caproliths are grayish or brownish and cylindrical in forms with pieces of bones and emit foul odour.

Chapter **2**

Disturbances of Growth

Atrophy

The term atrophy (L. or G.= not + to nourish) refers to a wasting away or a condition of diminution in bulk of tissue or organ due to lack of proper nutrition. The reduction in bulk of tissue can result from a decrease in size of the functioning cells of an organ or a part of the organ. This is called quantitative atrophy. An actual disappearance of component cells of an organ can also cause atrophy known as numerical atrophy of tissue. If there is a balancing increase of connective tissue in an atrophied organ, the reduction in the mass of the organ cannot be seen grossly.

Gross appearance

1. Decrease in size and weight of the affected part
2. Decrease in consistency and deepening of the color and reduction in the amount of blood.
3. In a paired organ, reduction in volume is clearly discernible on comparison of one organ with its fellow organ and is also noticed in the cases of lungs or kidneys etc.

Microscopic appearance

1. Decrease in both size and number of cells. Cell borders are distinct with deep staining of the cells.
2. Nuclei are smaller and more basophilic.
3. Presence of wrinkled or undulating capsule of an organ.
4. Non-atrophied tissue (for example, trabeculae in the spleen) may be too large and too numerous and account for too

much approximation of such tissues.

5. In muscles, sarcolemma and endomysium look like fibrous tissue and sarcoplasm may be narrower or may even disappear.

CAUSES

1. Deficient nutrition

This can cause following kinds of atrophies :

a. General atrophy
b. Local atrophy
c. Pressure atrophy

(a) General atrophy

It causes are as under :

1. Chronic starvation or malnutrition. An animal may be deprived of food due to some obstruction in the stomach interfering with passage of food. Some lesions or disease in the stomach may disable the animal to digest the ingested food. In starvation or state of inability of the animal to take food as usual, the following changes are seen :

1. Weakness and nervous symptoms.
2. Ketosis due to accumulation of substances like ketone bodies and beta oxybutyric acid etc. in blood from incomplete oxidation of fat.
3. Coma and death may follow.
4. Wasting of adipose tissue, stored glycogen and also of organs like liver, spleen, heart and kidneys etc.
5. Comparatively little change in skeletal and central nervous system.

(b) Local atrophy

Some obstruction of arterial flow may result in the local atrophy of the organ. Atrophy of heart muscle results from narrowing of the coronary artery. Presence of wasting, serous atrophy of fat, anemia and weakness in the animal's body refers to a condition called cachexia or state of inanition.

(c) Pressure atrophy

It also results from factors like deficient nutrition. When pressure is more marked on the cells, they show atrophy. Pressure from tumours or aneurysms on an organ or part of an organ can cause its atrophy. Atrophy due to pressure in the kidneys and liver is seen in case of hydronephrosis and amyloid infiltration in the liver respectively.

2. Disuse or diminished function

Cells are prone to atrophy due to lack of exercise or work. In disuse atrophy, the organs do not receive physiologic stimulation. This kind of atrophy is seen in extremities of patients confined to a cot for a prolonged period or in a rigid splinting of a limb. Ankylosis causes atrophy of the muscles.

3. Nerve lesions

It is essential for the organs of the body to receive proper nerve stimulation. If the reflex arc is disrupted or nerve supply to the organ, say, muscle is cut off or paralysed, muscles show atrophies. In poliomyelitis in man and monkey, muscle atrophy is seen.

4. Diseases of the endocrine glands

Hair follicles and sebaceous glands in skin can under go atrophy due to thyroid deficiency. In hypopituitarism, atrophy of the teats is seen.

5. Toxins

Atrophy or marked emaciation is found in cases of protracted fever, parasitic diseases and neoplastic conditions (e.g. cancer) due to toxic factors.

Physiologic atrophy

It is found in man and animals (e.g. resorption of tadpole's tail and involution of thymus etc.). Involution of uterus and mammary glands in females is another example following an end of pregnancy and lactation respectively.

SOME SPECIAL KINDS OF ATROPHY

1. Serous atrophy of fat

In this, fat cells in the pericardium and bone marrow rapidly disappear and their interstices get filled with fluid. The affected tissue looks like jelly as seen in the cases of Johne's disease and parasitic diseases.

2. Fatty atrophy

It is found in thymus in which missing cells are replaced by adipose tissue.

3. Fibrous atrophy or Scirrhous atrophy

Pre-existing cells or essential cells die or shrink and their places are taken up by proliferation of fibrous tissue.

Pigment or brown atrophy

This is seen in old age due to lack of nutrition or in the presence of some wasting condition. Pigments of wear and tear are seen near the poles of nuclei of the affected organs (e.g. heart) due to age or debilitating factors in animals.

HYPERTROPHY

Hypertrophy (Gr. =over + nourishment) means an increase in the volume or mass of an organ due to increase in size or number of its functioning cells. The term cellular hypertrophy means an increase in the size of its constituent cells where as hyperplasia (Gr. = over + formation) refers to an abnormal increase in the number of the constituent cells of an organ. Both cellular hypertrophy and hyperplasia result in the organ hypertrophy (i.e. its enlargement). Skeletal muscles and cardiac muscles have lost the power of mitosis and are able to undergo only pure hypertrophy. However, some limited cell division can be noticed in muscles. In other words, increase in either size or number of functioning cells will ultimately make an increase in volume of an organ (i.e. hypertrophy)

Gross appearance

The affected organs are larger than the normal cells.

Microscopic appearance

It looks like normal tissue but the cells are increased in number.

Classification

Hypertrophy can be classified as given below :

1. Physiological hypertrophy
2. Pathological hypertrophy

1. Physiological hypertrophy

It is seen in pregnant uterus and lactating mammary glands, repeated exercise of the muscle can lead to physiologic compensatory hypertrophy.

Pathological hypertrophy

The two types of pathological hypertrophy are as follows :

1. Compensatory hypertrophy (Adaptive hypertrophy)
2. Hormonal hypertrophy

1. Compensatory hypertrophy

It arises due to an impaired functioning of an organ or part of an organ system. There is an increased demand of work made upon the particular organ or tissue. If one kidney is lost, the other kidney undergoes hypertrophy and dose more work than usual for compensating the loss of work done by the fellow destroyed kidney. This is seen in organs like heart, stomach and intestine etc.

If there is a valvular disease or hypertension, heart shows adaptive ability and is increased in size as a result of an increased work, stress and strain thrown on the heart itself. Hypertrophy in heart may be confined to one or more chambers according to the site of obstruction. Stenosis of the pylorus of the stomach leads to hypertrophy of the gastric wall musculature. Obstruction in organs like intestine, ureter, urethra, urinary bladder can also lead to hypertrophy in the affected parts above the obstructed site.

2. Hormonal hypertrophy

It may be also physiologic type as seen in the testes of birds during mating season and mammary glands with the approach of lactation. Both the testes and mammary glands are enlarged. Organs show hypertrophy due to over production of hormones in body of an organism.

Causes

1. Increased resistance to the flow of blood. Right ventricle is enlarged due to pulmonary diseases and mitral stenosis.
2. Resistance offered to the passage of food, e.g. hypertrophy of the intestine or oesophagus above the point of stricture or obstruction.
3. Resistance to the flow of urine. This is caused by calculi or enlarged prostate.
4. Intermittent pressure on the skin. A badly fitting harness causes hypertrophy in the skin of a horse.
5. Diseases of the endocrine glands.

Hypertrophy of the organs of the body is seen in acromegaly and giantism etc.

6. Nerve lesions and other factors.

Hypertrophy can be caused by nerve lesions, increased nutrition, and increased blood supply. Sometimes, there is no increase in the essential or component cells of an organ but increase is seen in the fat or connective tissue elements. Such hypertrophy of non-essential tissue gives rise to condition of pseudo-hypertrophy or false hypertrophy.

Results

1. Hypertrophy may be followed by atrophy, in case, the affected organs work under more adverse conditions like increased strain or poor nutrition etc.
2. Hypertrophy restores the normal balance.
3. Hypertrophy may cause distortion in valves of heart with consequent heart failure.

HYPERPLASIA

The term hyperplasia (Gr. = over +formation) merely refers to an abnormal multiplication of cells in an organ in response to functional demands or different stimuli. It causes enlargement (i.e. hypertrophy) of the organ. Persisting callus is an example of hyperplasia.

Gross appearance

Enlargement of the organ is the only recognizable change to the naked eye.

Microscopic appearance

The cells are alike normal cells. However in glands, there is an increase in height of the glandular or acinar epithelium. This kind of hyperplasia is seen in thyroid and prostate. Huge acini (cystic glandular hyperplasia) can be seen in edometrium or mammary gland. Acanthosis is a hyperplasia of the stratum spinosum leading to thick prickle cell layer. Thickening of the cornfield layer (strum corneum) is called hyperkeratosis. In hepatic cirrhosis, hyperplasia of the liver cells may give rise to nodules of proliferating cells.

Hyperplasia can be divided into two types as given below :

1. Physiologic hyperplasia

If one kidney is removed, the other opposite kidney shows physiologic hyperplasia. Loss of blood causes physiologic response called erythroid hyperplasia (i.e. regenerative hyperplasia). The causes of hyperplasia include bacterial infection (lyphoid hyperplasia) protozol infection (hyperplasia of the bile duct in hepatic coccidiosis in rabbit), viral infection (epithelial or mesenchymal hyperplasia in pox diseases) and chronic irritation (epithelial hyperplasia in corns.)

2. Hormonal hyperplasia

It may be physiologic or pathologic. Mammary glands undergo physiologic hyperplasia with puberty and pregnancy. Due to malfunction of ovary, there is cystic glandular hyperplasia in these organs. Hyperplastic goiter is another example of pathologic hyperplasia.

Effects

1. Hyperplasia may progress to neoplasia due to procurement of power for uncontrolled proliferation by the cells.
2. Hyperplasia shows regression with withdrawal of the stimulus or irritant.

APLASIA (Agenesis)

It denotes the complete absence of an organ in the body because of deficient development of tissues. One notices absence of kidneys or uterus in the body. Thymus does not develop due to a defective gene in mice. Poisons like thalidomide produces amelia i.e an absence of limbs. *Veratum californicum* causes aplasia of palate in animals following its ingestion .

Hypoplasia

It is a developmental anomaly marked by a failure of an organ or a part of an organ due to defects in genes, hormonal deficiencies and certain poisons. It arises from incomplete or arrested development of organs.

Anaplasia

It refers to an undifferentiated state of cells i.e. a presence of an atypical primitive embryonic feature of cells. This condition is noticed in malignant tumours. Dedifferentiation of matured cells or tissues leads to reversion of primitive state in them. Disturbances in the normal constant architectural pattern of adult tissues, dis- organization of tissues, loss of polarity of cells and usual features of nuclei, hyperchromatism and mitosis are very noteworthy features of cellular anaplasia.

METAPLASIA

Metaplasia (Gr. = after + to form) refers to changing or conversion of one kind of tissue into another type. Columnar epithelium can change into stratified epithelium or fibrous tissue can be changed in to a bone or cartilage. Such changes, thus, happen within certain histologic limits. In other words, an epithelial cell cannot be turned into a connective tissue cell. Gallstones can cause metaplasia of the lining epithelial cells of the gall bladder.Vitamin. A deficiency can also cause metaplasia in epithelial elements. In cattle, bony spicules have been found in alveolar septa of the lungs. Bony or cartilaginous metaplasia is found in adenocarcinoma or mixed tumours of mammary glands in bitches. Sometimes, scars of the abdominal wounds show bony layers.

Significance

1. Protective purpose
2. Demand for a different kind of function
3. An almost irreversible change

Cellular aging

All cells undergo a fixed number of divisions and reach terminally a non- dividing state called cellular senescence (i.e. an old age in other words). Telomeres are repeated sequences of DNA (TTA GGG) which compose the linear ends of chromosomes. These are important in length determination of chromosomes. It is the active telomerase which adds repeated sequences to the ends of chromosomes. The regulatory proteins, in turn, suppress the telomerase activity. The sequences of DNA consist of specialized ribonucluoprotein (RNP) i.e. telomerase. The enzyme telomerase stabilises the telomere length by adding repeated sequences of DNA to the ends of chromosomes. In short, telomeres elongation is limited by regulating proteins and each cell division is marked by some shortening of structures e.g. telomeres located at the ends of chromosomes. No telomerase activity is seen in normal mature somatic cells and telomeres shorten until senescence sets in. Telomerase activity is shown by germ cells, stem cells and abnormal cells like malignant neoplastic cells. In cancer cells, telomerase activation inactivates the telomeric clock that is known to restrict the proliferative capacity of normal somatic cells. Shortened telomeres are seen in normal somatic cells and indicate about the growth arrest in cells. The cancer cells become immortal cells due to reactivation. The telomeres do not shortern in malignant cells and telomeres elongation is an important step in neoplasia (i.e. tumour formation.). Clock genes are reported in the lower forms of life. CLK-1, a gene of *nematode Caenorhaditis elegans* seems to alter its growth rate and times of developmental processes. The life span of a species of animals seems to be limited by fixed total consumption over a life time. The amount of oxidative damage by free radicals influence the phenomenon of senescence. Reduction of oxidative damage by restriction of caloric intakes in mammal prolongs their maximal life span. Antioxidative enzymes like superoxide dismutase (SOD) and catalase increase the life span in transgenic form of Drosophila. Repair of DNA damage by DNA repair enzymes influences cell age process of mammals. DNA damage is noticed in aging cells. Reactive oxygen metabolites like hydrogen peroxide (H_2O_2), superoxide anion (O2-) and nitric oxide (NO) are formed due to aerobic metabolism. Antioxidant mechanism inactivates these reactive oxygen species which hasten the aging processes. Reversible and irreversible changes occur due to interactions between the macromolecules and the reactive oxygen species. Aging cells and certain diseases are caused by acceleration of irreversible oxidative damage. Vitamin E, vitamin C, urate, glutothione superoxide dismutase (SOD) are antioxidant defence mechanisms. Cell aging is influenced by a balance between cumulative metabolic damage and response to that damage. Repair of DNA damages enhances the life span of animals. Restriction of oxidative free radical damage to proteins, lipids and DNA controls the processes of cell damage and the premature aging of cells in some diseases.

To sum up, the following facts are noteworthy of mention:

1. Telomeres (indicating chromosome ends) in chromosome shorten with each cell division. Telomere shortening arises from incomplete replication of chromosome ends division. Shortening of telomeres beyond a certain point is followed by loss of telomere function, chromosome, fusion and cell death.
2. Active telomerase prevents telomeres shortening in certain cells e.g. germ cells.
3. Most matured cells lack telomeres.
4. Introduction of the enzyme telomerase into the normal cells causes an extension of cellular life span. Regulatory proteins repress the activity of the enzyme telomerase as a result, telomere elongation is restricted.
5. Loss of telomeres is associated with loss of replication activity of cells. In other words, it indicates a beginning of the old age.
6. Telomerase activity is noticed in a large number of neoplasms e.g. cancer cells.
7. Telomere shortening in cells is considered to be a tumor suppressive mechanism of cells.
8. Proteins inhibiting cell growth cycle are over expressed in senescent cells.
9. Macromolecules (DNA, RNA, lipids and proteins) are oxidized by free radicals.

The oxidized macromolecules undergo irreversible molecular damage leading to cell aging. Nitric oxide also impairs cellular function and causes physiologic attrition. All the aforesaid factors lead to cell aging in the body.

Chapter **3**

Development Abnormalities or Distrbances

Tetratology refers to the developmental disturbances in an individual during embryonic or foetal life. Many malformations or abnormalities are frequently noticed in the new born or aborted fetuses of the animals. Anomaly is a developmental disturbance found in an organ or a part of an organ. Many of these anomalies are hereditary or heritable defects. Some of these defects are due to lethal genes which even cause death of the foetus or the new born. Some animals still survive with some abnormalities in their bodies and may remain of some economic importance to their owners. In general, dysontology deals with the study of developmental abnormatilies in the bodies of organisms. Atresia ani in calves arises from failure of the overlying skin to disappear from the anal opening. As a result calves (bovines) are unable to defaecate and may die ultimately.

Malformations can be divided into two types :

1. Heritable malformations
2. Non-heritable malformations

1. Heritable malformations

A heritable malformation due to lethal factor is noticed in Dexter cattle (an English breed with short legs). These animals do not breed true and Dexter, Kerry (normal) and bull dog calves are produced by them. Bull dogs are still born and show achondroplasia (recognized by a vaulted skull, split upper lip or protruded lower jaw and swollen tongue). Achondroplasia (*Chondrodystrophia foetalis*) has also been noticed in Telemark breed in Norway. Dwarfs due to a lethal gene are noticed in beef breed of cattle (Hereford) and these animals rarely reach a reproductive stage. Shorters or dwarfs which show peculiarity in breathing, stand with lowered head and neck and are also diabetic. In horses, a lethal character causes complete closure of ascending colon in the region of pelvic flexure. This state is known as atresia coli and has been noticed in white breed of dogs, white cats and other white animals.

Haemophilia is a heritable blood defect which is transmitted through females and the disease is due to a sex linked recessive factor or gene. Half of the males, produced by females which are carriers, are bleeders or haemophiliacs. Bleeders have prolonged clotting times. Haemophilia and haemophilia B arise from insufficient production of thromboplastin due to lack or decrease of anti haemophilic globulin (factor VIII) and plasma thromboplastic component (factor IX) respectively in the plasma of individuals. In certain pure bred dogs, haematocysts (blood cysts) are formed due to haemophilia. A heritable transmissible defect in blood coagulation has been seen in swine. Congenital porphyria (a heritable defect in the haemoglobin formation) has been noticed in cattle. The enzyme system responsible for the conversion of porphyrin to haeme is defective and the porphyrin laden red cells are haemolysed and the pigment porphyrins released from the cells are found in plasma and can also be deposited in the dentine of teeth i.e. the condition of pink took (congenital porphyria) is produced in cattle. Prophyrins can be detected in the urine of the patients (porphyrinuria).

2. Non-heritable malformations

The malformations for different kinds of defects or premature births are seen due to several nongenetical factors like aminoacids or injection of certain toxic material like selenium. Some of these factors are as follows :

1. Agenesia

It is caused by incomplete and imperfect development of an organ or part.

2. Aplasia

This refers to a complete absense or part organ or part in the body e.g. kidney and uterus etc.

3. Acrania

In this, there is an absence of most or all of the bones of the cranium.

4. Anencephalia

It refers to the absence of the brain in most animal species.

5. Hypocephalia

In this, there is an incomplete development of the brain.

6. Hemicrania

This refers to an absence of half of the head.

7. Exencephalia

In this, there is defective skull with brain exposed or extruded.

8. Arhinencephalia

It refers to an absence or rudimentary enlargement of the olfactory lobe. There is also lack of development of external olfactory organs.

9. Agnathia

It means absence of the lower jaw.

10. Abrachia

It refers to the absence of fore limbs.

11. Abrahiocephalia

It means the absence of fore limb and head.

12. Adactylia

It refers to an absence of digits.

All the aforesaid malformations are produced due to arrest of development of the organs or parts.

Development of fissures on the median line of the head, thorax and abdomen also give rise to malformation as given below :

Cranioschisis

It refers to failure of cranial fusion and has been seen in pigs and gives rise to herniation of meninges, (meningoencephalocele).

Cheiloschisis

It is also called hair lip.

Palatoschisis

It is called cleft palate. Faulty development of the maxillary process derived from the 1st visceral arch gives rise to hair lip and cleft palate.

Rachischisis (Schistorrachis)

It is also spinabifida

A congenital defect characterized by failure of closure of the dorsal part of the spinal column and may extend from sacral region to the occipital bone. The upper spinal canal may have a covering of skin.

Schistthorax (thorax or sternum)

Schistosomus (abdomen)

Fusion of paired organs gives rise to foetal anomalies. These are as follows:

Cyclopia

In this, one notices one eye or fused eyes in single orbital fossa. *Veratrumcalifornicum* produces partial or complete cyclopia on its consumption by sheep.

Excess of development of the organs can give rise to anomalies. It indicates a congenital hypertrophy. Anomalies also arise from increase in the number of some parts (say, ears) of the body.

The following are the examples :

1. Polyotis (ears)
2. Polydontia (teeth)
3. Polymelia (limbs)
4. Polyodactylia (digits)
5. Polythelia (teats)
6. Polymastia (mammary gland)

Displacements during development of the embryo or foetuses gives rise to malformations due to displacement of the organs are as under:

1. Dextrocordia

It refers to the transportation of the heart to the right side of the body.

2. Ectopia cordis cervicalis

It refers to displacement of the heart to the neck. It has been found in cattle.

3. Ectopia cordis

In this, the heart lies subcutaneously outside the thorax.

Anomalies arising from displacement of the tissues are as follows:

(i) Teratomas

These are neoplasms arising from foetal residues or rests misplaced or displaced within individual. Dermoid cysts and dentigerous cysts are examples of teratomas.

Persistent foetal structures

Many foetal structures disappear during intrauterine life or soon after birth. The failure of disappearance of such structures (e.g. ductus arteriosus, urachus Wolfian and vitelline ducts) creates a pathological state of persistence of such structures.

Persistent ductus arteriosus

It arises from failure of the development of aortic septum (i.e. presence of a common aorta and pulmonary artery. When there is a partial failure of the development of the aortic septum, there is a communication between the aorta and pulmonary artery. Such patients suffer from cyanosis and death occurs ultimately. The ductus arteriosus carries blood from the pulmonary artery to the aorta in the foetus and appears as a fibrous band in the postnatal life connecting the pulmonary artery to the aortic arch. When the ductus arteriosus is not obliterated, a patent ductus arteriosus exists. The shunt is present and the blood is shunted from pulmonary artery to systemic circulation.

Patent foramen ovale (i.e openness of the foramen ovale).

The atria communicate with each other through open foramen ovale in the foetuses. If it is left as a fosa after birth, it is seen in the form of diverticulm in the septal wall. When the foramen ovale fails to close entirely, the state of patent foramen ovale is created. It is an atrial defect.

Pervious urachus

When the foetal urachus fails to close, previous urachus is produced. A patient can get rid of it by surgical step. Infection can spread from the umbilicus (navel) to the urinary bladder.

Hermaphrodites

Fusion of sexual organs leads to the state of hermaphrodites and pseudohermaprodites.

A hermaphrodite is an individual who has both the testicular and ovarian tissues. This is a true from of hermaphrodite. But in a false hermaphrodite, presence of either of ovary or testes is accompanied by the simultaneous presence of secondary sexual characters of the opposite sex. False hermaphrodites have been found in goats, pigs, cattle and other animals and both types are not usually fertile.

A freemartin is known as a female calf co-twin to a bull. Such inter-sexes are found in cattle and are mostly sterile. The female of heterosexual twins is a freemartin in which female genital organs fail to develop and vestigial male organs can be found due to dominant male hormones produced by the male twin. The genital organs do not develop in females and vagina and vulva are present with enlarged clitoris. There is also retarded growth of the mammary glands. A heifer is born twin to a bull normal in 7 percent of cases. In short, there is an anastomosis of the circulation of the fetuses owing to the fusions of chorions in twin pregnancy and male hormones are carried to the female by the circulation adversely affecting the development of female genitalia.

MONSTERS

A monster is born due to disturbance of development which is characterized by great or abnormal distortion of an individual. There is duplication of all or most of the organs and other parts in an animal which develops from a single ovum.

Monsters belong to the same sex and develop with a single chorion. Incomplete twining gives rise to monsters and can be produced in birds and sea minnow by subjecting the embryos to adverse stimulations in the late cleavage stages.

Twins can be separated into two classes :

1. Twins entirely separate
2. Twins united

Twins entirely separate

Such twins develop in a single chorion. One of the twins is well developed whereas the other is malformed (acardius). Heart, lungs and trunk show arrested development in acardius which may show absence of head (acephalus) or limbs of trunk (acormus).

Twins united

There can three categories of twinings which are as follows :

(i) Anterior twining (ii) posterior twining (iii) twining almost complete.

(I) ANTERIOR TWINNING

It is united in the pelvic region with bodies side by side.

1. PYGOPAGUS

It is united in the pelvic region with bodies at an obtuse angle.

2. ISCHIOPAGUS

It is united in the pelvic region with bodies at an obtuse angle.

3. DICEPHALUS

In this, there are two separate heads and doubling may be noticed in the neck, thorax and trunk. According to the number of anterior and posterior extremities, dicephalus may be known as tetrapus, tripus, dipus, tetrabrachius, tribrachius and dibrachius.

4. DIPROSOPUS

In this, there is doubling in the cephalic region without complete separation of the heads. The face is found doubled. Depending on the number of eyes, diprosopus can be divided into the following types. (1) Tetrophthalamus (2) Triophthalamus (3) Diophthalmus.

According to numbers of ears, the kinds of twins are tetrotus, triatus and diotus. Monostomus and distomus types depend on the number of mouths in twins. Number of fore limbs can designate them as tribrachius and dibrachius etc.

POSTERIOR TWINING

In this, the posterior part is double and the anterior part exists singly. In human monsters, the position of the twin will be side to side (lateral), back to back (dorsal) and abdomen to abdomen (ventral) and posterior twining gives rise to following 3 types :

1. CRANIOPAGUS

In this, brains are usually separate and bodies are found at an acute angle. Depending upon the place of attachment, the forms or types are called parietalis, frontalis and occipitalis.

2. CEPHALOTHORACOPAGUS

In this, there is union of head and thorax.

3. DIPYGUS

In this, there is a doubling of the posterior extremities and posterior part of the body. Dipygus may also exist as dibrachius or tetrabrachius.

TWINING ALMOST COMPLETE

It is a twin in which joining of the pair can be seen in the region of thorax or abdominal region. There can be duplication of the whole trunk of the anterior or posterior extremities (the parallel ventral arrangement of the foetuses).

1. THORACOPAGUS

In this a, twin is united by thorax. Designation of the monsters is dependent on the number of fore and hind extremities.

2. PROSOPOTHORACOPAGUS

In this, there is union of the thorax; designation of the monsters is dependent on the number of fore and hind extremities.

3. RACHIPAGUS

This twin shows union of the thoracic and lumbar processes of the spinal column.

In monsters, the parasite (marked by one of the pair smaller and less developed than the other) is partially embedded in the autosite (normally developed one).

Some information about other developmental abnormalities are given in Table-4

Table 4: Some developmental abnormalities (defects)

Defects	Features of these defects
1. Ectopia cordis	A term indicating location of the heart outside the chest cavity.
2. Cardiac hypoplasia	A small heart not proportionate to the weight and size of the body
3. Horseshoe kidney (Sigmoid kidney)	A term applied to fusion of the kidneys resulting in formation of such shape of kidneys .
4. Congenital cystic kidneys	Congenital cysts in the kidneys probably due to failure of the fusion of the glomeruli with the collecting tubules.
5. Intersexes or free martins (an example of congenital defects due to hormonal imbalance).	A female calf co-twin to a bull.
6. Anencephaly	Marked by partial or complete absence of the brain in the body.
7. Acrania	Indicating a partial or complete absence of bones of the cranium.
8. Meningocele	Pertaining to protrusion of the membranes (meninges) through imperfectly developed cranium.
9. Encephalocele	Relating to the presence of the brain substance in the protruding membranous sac through the defective cranium
10. Microcephalus	Pertaining to presence of small head in the body.
11. Microencephalus	A brain smaller than normal.
12. Hernia	A defect in the abdominal wall or diaphragm causing the condition of hernia e.g. protrusion of the viscus through natural or accidental orifice (opening).

Chapter 4

Disturbances of Circulation

Disease processes like hyperaemia, haemorrhage, oedema, thrombosis, embolism and infarction etc. are examples of important circulatory disturbances.

Hyperaemia (Gr. = over + blood) and congestion (L. = heap together) refer to an excessive accumulation of blood in the vessels of tissues or organs of the body. It is divided into two groups, namely (1) active hyperaemia and (2) passive hyperaemia. It is characterized by an increased inflow of blood to a part due to active dilatation of both the arterioles and capillaries. The dilatation of these vessels is partly due to nerve stimuli and partly due to histamine like substance or H-substance causing contraction of the arterioles and dilatation of the capillary in the inflamed tissues.

The basic necessities for normal activities in cells and tissues include an adequate delivery of oxygen and normal fluid balance in organs. Abnormalities in either blood supply or fluid balance in tissues produce morbidity and mortality in organisms. Pulmonary oedema arises from ischemia, valvular diseases and circulatory disturbances like congestion and thrombosis etc.

Active hyperaemia or Hyperemia

The increase in the blood supply to an organ (hyperaemia) occurs due to dilatation of the arterioles and capillaries. It is divided into physiological and pathological hyperaemias. Increased blood supply into the gastric mucosa during digestion and the uterus during pregnancy are examples of the physiological hypeaemia.

Active hyperaemia may be caused by :

1. Noxious agents (e.g. oil of mustard and turpentine) acting on the vascular walls in acute inflammation.
2. Paralysis of the vasoconstrictors.
3. Stimulation of the vasodilators.
4. Increased collateral circulation (i.e. collateral hyperaemia) due to interference with blood flow through a neighbouring vessel.
5. Physical factors, e.g. heat, and light.
6. Mechanical irritation like friction.

Gross appearance

The affected tissue is discoloured or bluish red and usually swollen. There may be some increase in temperature of the tissue.

Microscopic appearance

The capillaries in the tissues are dilated and engorged with blood or numerous red cells and the extracapillary tissues are permeated with fluid and red cells. In health, the lumen of a functioning capillary is slightly wider than that of a red cell and only one red cell is usually accommodated inside the capillary lumen.

PASSIVE HYPERAEMIA OR CONGESTION

In this, there is a decreased outflow of blood on the venous side of circulation. Some obstruction to the normal outflow of blood produces passive hyperaemia.

Passive hyperaemia is divided into (1) general passive hyperaemia and (2) local passive hyperaemia.

General passive hyperaemia

It may be acute or chronic.

Acute general passive hyperaemia

In this, an excess of blood is found in most tissues or organs due to failure or obstruction in the areas of the body through which the flow of all blood is a must and such areas or organs are heart and lungs in the body. The condition is an acute hyperemia which develops rapidly and lasts over a short period and it may be fatal and has been noticed in race horses, hunters and grey-hounds, etc. Severe exercise after a prolonged rest and inefficient cardiac action are the causes of death in these animals. The affected organs are deep, red and bluish red and when incised, the blood flows freely on the cut surfaces of such organs (for example, lungs). In case of heart failure or a failing heart, acute general passive hyperemia develops. The large veins in the thorax and abdomen are dark and engorged with blood and liver and lungs are congested and veins in the mesentery and intestine undergo hyperemia. Such important gross changes are visible in the dead bodies postmortem.

Chronic general passive hyperaemia

It arises from a gradual, long standing cardic failure from chronic valvular diseases. Obstruction to the flow of blood in the lungs as noticed in narrowing of the vessels in emphysema and diffuse fibrosis also causes general passive hyperaemia. Stenosis from valvular endocarditis or thrombosis may also induce such chronic general congestive state.

Local passive hyperaemia

It is noticed in an organ or part due to an obstruction to the venous return from these areas. It may be acute or chronic.

Acute local passive hyperaemia

Its main causes are:

1. Pressure on a vessel from outside by a tight bandage, tumour or pregnant uterus, hernia, volvulus or haematoma etc.

Obstruction within the vessel (e.g. a thrombus), leads to marked serious changes in the organ in absencc of collateral circulation. The affected organ shows accommodation or adaptation in the case of gradual application of obstruction to the flow of blood in the vessel and availability of collateral circulation. If the obstruction to blood flow in the vessel is rapid and complete, very extensive damage like necrosis and other damages occur in the affected organs.

Pathological changes in passive hyperaemia or congestion

The affected parts in passive hyperaemia are cold, engorged with blood, swollen, purple in colour (cyanosis) and devoid of sensation. Later, these organs may be firmer than normal to constitute what is called cyanotic induration. Induration (hardness) in the affected tissue is due to an increase of connective tissue elements chiefly in the wall of capillaries and venules. The veins may be extremely dilated, congested and tortuous in equines and bovines due to pressure of the foetus on them (varicosity).

There is an engorgement, dilatation of capillaries and venules escape of fluid into the extravascular tissues, extra-vasation of red cells and acute necrosis of tissue (infarction) in complete occlusion of the vessel.

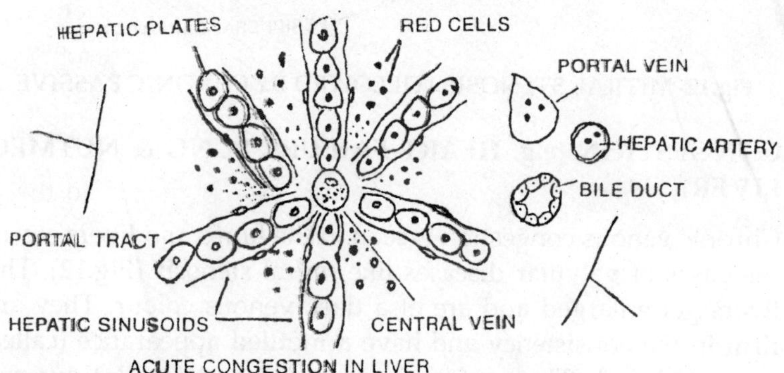

Fig. 11

In acute general passive congestion, due to failing heart (i.e. in congestive heart failure), the vena cava are large veins of the abdominal and the thoracic cavities and dark and engorged with blood. The veins in intestinal walls are dark, prominent and the livers and lungs may contain more blood than usual.

The sinusoids and capillaries in livers are dilated (Fig. 11) but empty due to oozing out of blood into the solution during the processes of cell staining. The central cells of the hepatic cords tend to disappear leaving behind a central area of blood stagnation.

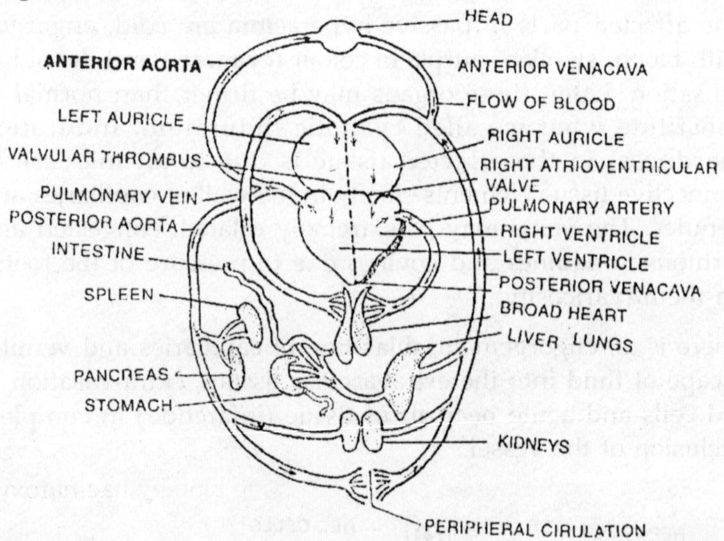

Fig.12- MITRAL STENOSIS FOLLOWED BY CHRONIC PASSIVE

CONGESTION (e.g. HEART FAILURE LUNG & NUTMEG LIVER)

Chronic venous congestion is seen in the lungs and livers etc. in the cases of valvular diseases like mitral stenosis (Fig.12). The livers get enlarged and are of a dark venous colour. They are firm in the consistency and have a mottled appearance (called nutmeg liver). The surface is granular and lobulations are distinct. There is a dark congested central zone surrounded by yellowish peripheral area in the lobules. Microscopically, the sinusoids are enlarged or distended and fibrous tissue supporting the endothelium is increased. The liver cells in the centre of the loubles show atrophy, degeneration and necrosis. In peripheral part of the lobules, the liver cells show fatty changes. Thus, the lobules are red at the centers but pale at the peripheries. Central veins and sinusoids are thickened due to proliferation of connective tissue. An increase in the connective elements in the walls of central veins is called central cirrhosis.

In chronic passive congestion of the lungs, the lungs are enlarged deep red or bluish red and become firm and brown in colour (brown induration of lungs). The alveolar vessels are dilated and the alveolar spaces contain red cells, fluid and macrophages laden with haemosiderin. These macrophages contaning haemosiderin are called heart failure cells.

To sum up, hyperemia means an active process in tissues marked by an increased flow of arterial oxygenated blood because of arterial dilation in the affected organs whereas congestion denotes a passive process arising from an impaired outflow of blood from an organ. Venous obstruction causes a passive congestion and the deoxygenated blood stagnated in the organ imparts a bluish red colour (cyanosis) to tissues . In infective hyperemia, the affected organ or part of an organ shows redder colour in the tissues. Chronic hypoxia is noticed in chronic passive congestion of long standing nature in an organ. Chronic pulmonary congestion is marked by an intraalveolar haemorrhage or septal oedema, thickened and fibrotic alvreolar septae and haemosiderin loaded macrophages (called heart failure cells). The term heart failure lung refers to lung revealing induration (hardness), brown discoloration and heart failure cells in the alveolar spaces of the lung. Central veins and sinusoids of the liver in chronic passive congestion are distended with blood. Degenerative changes (e.g. fatty changes) are noticed in the peripheral hepatocytes of the hepatic lobules. In short, red brown centres and pail peripheral areas are noticed in lobules in chronic passive congestion. Grossly, the liver resembles a nutmeg. And as such, the liver is designated as a nutmeg liver. Necrotic changes in hepatocytes and fibrotic changes in the central veins of the hepatic lobule are very important histopathlogical features.

OEDEMA (Dropsy)

The term oedema (Gr. = swelling) refers to an excessive amount of fluid in the cells and intercellular spaces of the body and effusion (L. = pouring out) means the process of its formation. Initially, the fluid is confined to the tissues, but, on being overburdened with it, it accumulates in the body cavities or depended part of the body and subcutaneous tissues. This state is called dropsy (Gr. = water) or anasarca (Gr. = throught out + flesh). Anasarca is characterized by fluid accumulation in subcutaneous tissues. The fluid that is formed in response to non- inflammatory causes is called transudate (L. = through + to sweat).

Excessive amount of fluid accumulation in the cells gives rise to a state of cellular oedema whereas an abnormal accumulation of fluid in the tissue spaces or body cavities refers to the commom term known as oedema or dropsy. Normally, the tissue spaces or body cavities contain some fluid and any excess of fluid in such areas is drained away by the lymphatics as lymph. In health, the capillary blood pressure at the arterial end of the capillary system is greater than osmotic pressure whereas at the venous end of the capillary net work, osmotic pressure is greater than the capillary blood pressure at the venous end of the capillary net work. This results in diffusion or escape of fluid at its arterial end of the capillary net work and reabsorption at the venous end. These repulsive and attractive forces at the arterial and venous ends of the capillary bed are so balanced or adjusted as to maintain the usual normal amount of the fluid in tissue spaces.

The chief causes of oedemas are :

1. Increased capillary blood pressure.
2. Decreased colloidal blood pressure.
3. Increased permeability of the capillary walls.
4. Increased colloidal osmotic pressure of the tissues.
5. Lymphatic obstruction.

Pathogenesis of oedema

In development of oedema, there is an increased transudation of fluid from vessels into tissues and retention of the fluid by the tissues. The impeded outflow of the fluid through the lymphatic vessels is conducive to development of oedema but presence of collateral lymph channels and out-flow of tissue fluid through the veins minimizes the adverse effects of lymphatic blockage. Oedema is due to disorders of water interchange between the blood and the tissues. Capillary blood pressure (35-40 mm Hg) is greater than colloidal osmotic pressure (25-30 mm Hg) at the arterial end of the capillary in mammals (say, human beings) and this favours an increased transudation of fluid from the blood vessels into tissues and the effused fluid is retained by the tissues. But at the venous end of the capillary bed, hydrostatic blood pressure (15-16 mm. Hg) is lower then the blood colloidal osmotic pressure (25-30 mm. Hg and this condition favours increased absorption or intake of fluid from tissues into the blood vessels. Thus, the fluid exchange is balanced between capillary bed and tissues in health (Fig. 12). Tissue colloids have the ability to bind water to swell and an increase in their concentration in the extravascular tissues helps development of oedema due to enhanced water attractive force in the tissues. Increased permeability of the capillary wall helps the pathogenesis of oedema. Increased capillary pressure and decreased osmotic pressure of the blood is opposed to each other in working directions. The former is an expulsive force and the latter is an attractive force and, these two opposed forces cause increased transudation of fluid. The capillary blood pressure in the venous part of the capillary bed may rise to 25-40 mm. Hg (instead of the normal 15-26 mm Hg.) and this change leads to increased filtration of fluid from the blood into the tissues. Increased capillary blood pressure with decelerated blood flow in organs of the body is noticed is cardiac insufficiency or obstruction of the veins. Proteins are of paramount importance in the development of oedema. There is high ability of tissue colloids to bind water. An increased concentration of tissue colloids (e.g. in the inflamed tissues) favours pathogenesis of oedema. A decrease in the concentration of proteins in the plasma from 5.5% of albumins to 2.5 gm percent may lead to oedema. A decrease in plasma proteins to 3% in dogs causes oedema of subcutaneous tissues. Increased blood pressure in the capillary alone or only decreased colloid osmotic pressure of plasma does not always lead to development of oedema. There are actually interactions between several other factors and the predominant factor in pathogenesis of oedema involves an inter-play of several factors. The endocrine glands like thyroid, hypophysis and aderenal glands play important role in the pathogenesis of oedema. Increased secretion of mineralo-corticoids causes retention of hydrophilic sodium in an organism. Aldosterone increases re-absorption of sodium into convoluted tubules of the kidneys and block its exeretion from body of the organisms. The fluid rich in proteins accumulates in the inflamed tissues. Transude and exudate differ from each other in the following respects (Fig. 13):

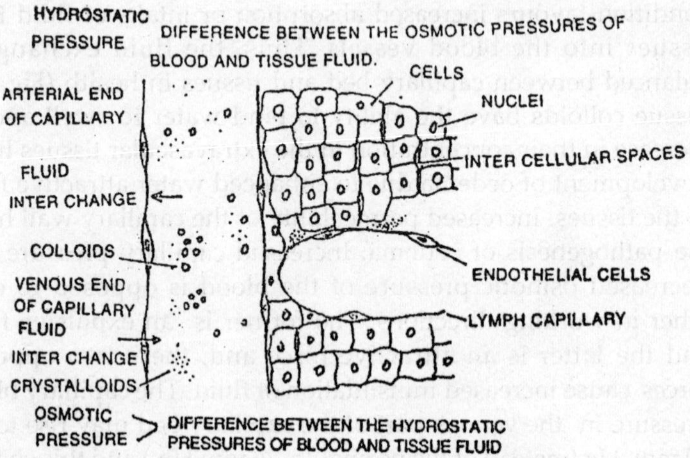

Fig. 13 FLUID INTERCHANGE BETWEEN CAPILLARY AND TISSUE

Transudate	Exudate
1. Low specific gravity (1006 to 1012), pale.	1. High specific gravity (1018) or more.
2. Low protein content (usually 0.3 percent)	2. High protein content (more than 4 per cent)
3. It contains very few cells cells, chiefly monocytes and lymphocytes	3. It contains large number of inflammatory cells like neutrophiles, bacteria and dead cells etc.

Different terms have been given to oedemas on the basis of different kinds of the tissues spaces or cavities involved. Oedema is noticed in various situations. It is found in different serous cavities and subcutaneous tissues etc. and has been designated by using different terms in view of their development sites as given below.

1. An accumulation of fluid is serous cavities means dropsy
2. An accumulation of fluid is subcutaneous tissue is called anasarca .
3. An accumulation of fluid in peritoneal cavity implies ascites.
4. Collection of effusion in pericardial cavity means hydropericadium.
5. An accumulation of fluid in thoracic cavity denotes hydrothorax.
6. An accumulation of fluid within the skull means hydrocephalus.
7. An accumulation of fluid in the sac of tunica virginals is called hydrocele.

Oedematous subcutaneous tissues are swollen, puffy and the tissues have moveable fluid and pit on pressure i.e formation of pits in the skin. On incision, fluid escapes from the oedematous tissues with temporary relief from an oedema.

Classification of oedemas

Oedemas can also be classified according to their aetiology into different types such as congestive, renal, cachectic, toxic and inflammatory types and some of them are as under :

1. Congestive or cardiac oedema

Congestive oedema arises from an obstruction of veins due to thrombi and emboli or from constriction by tumours or ligation. Disturbances in cardiac activity can cause oedema through a rise in blood pressure with consequent flow of fluid from the blood into the tissues. In congestive or cardiac oedema, there is also increased permeability of the capillaries due to distension and disturbed nutrition of their walls. In short, the factor of congestion is the main reason for this development.

The fluid can accumulate in the limbs and dependent tissues under the influence of gravity. Congestive oedema is found in cases of mitral stenosis, fibrosis and emphysema of lungs. Cirrhosis and echinococcosis in the liver obstructs portal circulation and ascites may follow portal congestion and hypoproteinaemia from liver damage.

2. Renal oedema

It arises in diseases of the kidneys partly due to retention of sodium in the tissues and partly due to decreased concentration of proteins in the blood (hypoproteinaemia). In glomerulonephritis, edema develops due to disturbances in the colloidal properties of the tissues and increased permeability of the vessels due to toxic substances. In acute renal oedema, there is an increased permeability due to toxins with resultant excessive transudation in the tissues. In chronic renal oedema, ther is an albuminuria. There is marked hypoprotinaemia and transudate has a low specific gravity and a low protein content.

3. Cachectic oedema or nutritional oedema

This is caused by chronic diseases (for example, wasting conditions, severe anaemia, malignant tumours and starvation etc.). In such oedema, there is an increased permeability of the capillary walls (due to malnutrition of the endothelium) and decreased concentration of proteins in the blood (due to insufficient intake by the organism). As a result, plasma filters more easily through the capillaries and less fluid is reabsorbed. Cachetic oedemas are found in distomatosis (liverfluke disease), parasitic pneumonia, cancer and tuberculosis.

4. Inflammatory oedema

In this type, there is an increased colloid osmotic pressure in the tissues and increased permeability of vascular walls. It is caused by insect bites (bee stings), anthrax, mustard gas, leuwisite, croton oil, chlorine and diphosgene etc. It also occurs in gas gangrene, black quarter, and in these infections, diffusion of toxins causes oedema in different tissues. Toxins and H substances are direct causes of oedema by producing changes in the capillary walls. (i.e. an increased capillary permeability)

5. Pulmonary oedema

Pulmonary oedema arises from inflammatory conditions or due to some inefficiency of left side of the heart. Inhalation of irritant vapours causes pulmonary oedema. It is also found in the lungs of horses suffering from African Horse Sickness. It causes hypoxemia, cardiac weekness, circulatory disturbances and death of the patients.

The out-comes of oedemas are:

1. When the cause is removed, the fluid accumulated in the tissue is reabsorbed with return to normalcy.
2. Loss of elasticity of the tissues in a protraected oedema
3. Dysfunction of the affected organs is found. In dropsy, there is compression of the surrounding organs or tissues. In pulmonary oedema, the fluid accumulated in the alveoli, considerably diminishes the respiratory surface of the lungs. The effect is almost similar to that of one as seen in the cases of drowning in water. Odema of lungs causes water logging in the pulmonary alveolar parenchyma leading to the state of anoxia or asphyxia.
4. Bacterial infection may set in oedematous tissues.
5. Death may follow.

To sum up, the movement of fluid between vascular and interstitial or intercellular spaces is governed by opposing effects (i.e. colloid osmotic pressure of the extravascular tissue proteins and colloid osmotic forces of blood proteins), vascular hydrostatic pressure, capillary hydrostatic blood pressure and lymphatic drainage of the exuded fluid as lymph from the blood vessels. There is nearly a balanced state in relation to any outflow of fluid into the interstitium from the arteriolar end of the blood channel i.e. microcirculation comprising of arterioles, capillaries and postcapillary venues and inflow of fluid at the venular end of the capillary bed. Normally there is a balanced state because of no net loss or gain of fluid from the capillary bed. Lymphatics remove any left over amount of the fluid from the intercellular spaces.

Oedema refers to an increased amount of interstitial fluid which is caused by the following factors:

1. Increase capillary blood pressure.
2. Decreased capillary osmotic pressure.
3. Increased colloid osmotic pressure of tissue proteins.
4. Lymphatic obstruction.

Any retention of sodium in renal diseases e.g. glomerulonephritis and acute renal failure produces oedema in the body in the inflamed tissues. There is a breakdown of exuded blood proteins from the capillary and other proteins resulting in an increased number of protein molecules and increased colloid pressure in the extravascular spaces. An escape of blood proteins into the extravascular spaces reduces the effectice osmotic pressure of the blood. Increased hydrostatic pressure is noticed in congestive heart failure, constrictive pericarditis and venous obstruction etc. Decreased plasma osmotic pressure, malnutrition, liver cirrhosis and glomerular nephritis causes oedema because of an escape of fluid from the capillary bed into the extravascular or intercellular spaces. Lymphatic obstruction arises from inflammatory conditions, neoplasms and parasites. Disturbances in the nervous system causes oedema. Administration of large amount of normal saline solution coupled with stimulation of the interceptors of the intestinal loop causes pulmonary oedema in rabbits. Cattle do not develop oedema from an excessive intake of only water or only sodium chloride (i.e. common salt oedema)

Haemorrhage

Haemorrhage (Gr. = blood + burst forth) means bleeding from the body and is caused by an escape of blood from blood vessels like arteries, veins and capillaries etc. External haemorrhage refers to haemorrhage (the escape of blood) on the external surface of the body whereas a haemorrhage into the body cavity or tissue is called internal haemorrhage.

Two kinds of haemorrhages are :

1. **Haemorrhage by rhexis i.e.** there is an escape of blood owing to rupture of the wall of blood vessels as seen in the cut or perforation of the arteries or heart.

2. **Haemorrhage by diapedesis**

This is a kind of haemorrhage in which the permeability of the capillaries is increased or minute imperfections develop in the wall of the capillaries or small blood vessels. Noxious agents like toxins or chemicals etc. increase the permeability of the capillaries to an extent of allowing the escape of the red cells (blood) outside their walls while the walls of the capillaries still remain intact. The other important causes of haemorrhage are:

1. Congenital defects in the muscles or coats of the blood vessels,
2. Defects in the internal elastic lamina.
3. Arteriosclerosis, and mucoid degeneration of media of artery.
4. Diseases, ulceration, accidental trauma (cuts, tears, bruises), stab wounds, gunshot wounds etc. in the walls of the blood vessels.
5. Increased intraluminal pressure due to hypertension.
6. Capillary and venous passive hyperemia.

All the aforesaid factors can produce rupture in the wall of the blood vessels resulting in haemorrhage.

Haemorrhages may be classified in various ways :

1. Hemorrhages classified according to the organs are cerebral, pulmonary, gastro-intestinal and uterine etc.
2. Haemorrhages are classified according to the sizes. Pinpoint haemorrhages of blood especially in the skin, mucous membranes etc. are called petechiae (L. = spots). These are found in some infections. Ecchymoses (Gr. = out +juice)

are still larger extravasations or haemorrhagic spots on a body surface or in the tissues. Numerous small haemorrhages can be found in the skin, subcutaneous tissues and occasionally internal organs and these are called purpuric spots (L. = purple).

Haematomas (Gr. = blood +tumor) are extra-vassations of sufficient degree to produce a sizeable swelling.

3. Haemorhages are also classified according to the sites of body or various situations.
   - (i) Epistaxis (Gr. = above + to drop) means the hemorrhage from the nose.
   - (ii) Haemoptysis (Gr. = blood + to spit) refers to the hemorrhage from the lungs.
   - (iii) Haematemesis (Gr. = blood + vomiting) means the vomiting of blood.
   - (iv) Haematuria (Gr. =blood + urine) is a passage of blood in the urine.
   - (v) Haemothorax (Gr. = blood +chest) refers to an accumulation of blood within pleural cavity.
   - (vi) Haemopericardium (Gr. = blood + pericardium) means a collection of blood in the pericardial cavity.
   - (vii) Apoplexy (Gr. = from + to strike) means the copious bleeding into an organ like brain.
   - (viii) Haematocolps (Gr. = blood + vagina) refers to collection of blood in the vagina.
   - (ix) Melaena refers to haemorrhage in the gut or intestinal tract discolouring the faeces .

Blood collection in the connective tissue or in the tissue of an organ (called haematoma) is seen in the ears of dogs and in the spleen of cattle, pigs and dogs.

Causes

1. Mechanical trauma (i.e. cutting in the wall of a blood vessel)
2. Necrosis or destruction of a vessel wall by an ulcer, a

spreading neoplasm and rupture from a vessel weakened from aneurysm or arteriosclerosis.

3. Toxic injury to capillary endothelium. Transient openings of punctiform sizes in the blood vessels give rise to petechiae or ecchymoses on serous and mucous membranes or even in the depth of the tissues. Toxins produced in hog cholera, anthrax, haemorhagic septicaemia, black leg and certain chemicals (arsenic) and plant poisons cause formation of petechiae in tissues due to development of imperfections in the vascular walls.

4. **Anoxia**

It is caused by direct suffocation, anaemia and produces petechiae and ecchymoses in the tissues.

5. Disorder of clotting mechanism can cause haemorrhage.

Effects

1. Hemorrhage in the brain results in the loss of vital functions.

2. Hemorrhages in pericardium interferes with cardiac action.

3. Hemorrhage in trachea and bronchi produces asphyxia.

4. Loss of more than 1/4 or 1/3 of the total blood of the body may be fatal. This causes drop in arterial blood pressure.

5. Thirst is increased.

6. Volume of the blood is replaced by withdrawal from intercellular spaces of tissues. Leucocytes are replaced in one or two weeks and erythrocytes in 6 weeks post haemorrhage.

THROMBOSIS

Thrombosis (Gr. = coagulation) refers to formation of blood clots in the blood vessels. (Fig. 14) These intravascular blood clots are called thrombi (Gr.= plug). Thrombus causes an obstruction to the circulation in the blood vessels and is found in any part of the heart, arteries, veins and capillaries. Clotting is due to involvement of two mechanisms, namely, (i) the intrinsic clotting mechanism and (iii) extrinsic clotting mechanism. There are several other factors like tissue thromboplastin (ii) Factor VII etc. in extrinsic system and factors like Hageman factor (xii), P.T.A. (xi) Christamus factor ix etc. in the intrinsic system. Inshort, prothrombin is converted into thrombin which acts on plasma protein fibrinogen to form fibrin. This conversion is helped by calcium ions and thrombokinase released from thrombocytes or tissues etc. Fibrin consists of a fine thread like net work entangling the formed elements (red cells and white cells etc.) of blood to form blood clot. A clot so formed in the vessels during life is different from the clot formed after death of an animal.

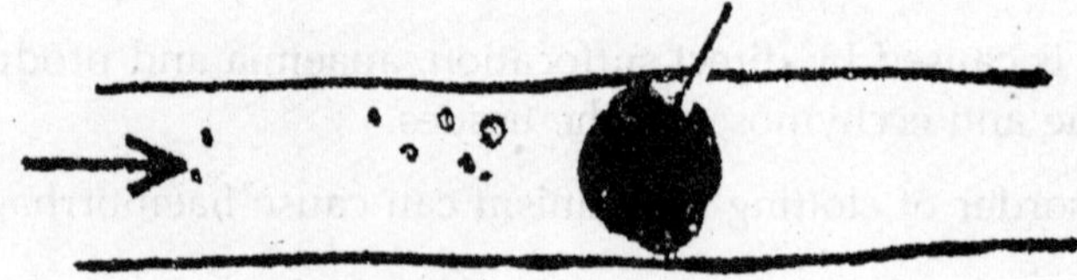

Fig– 14 THROMBUS

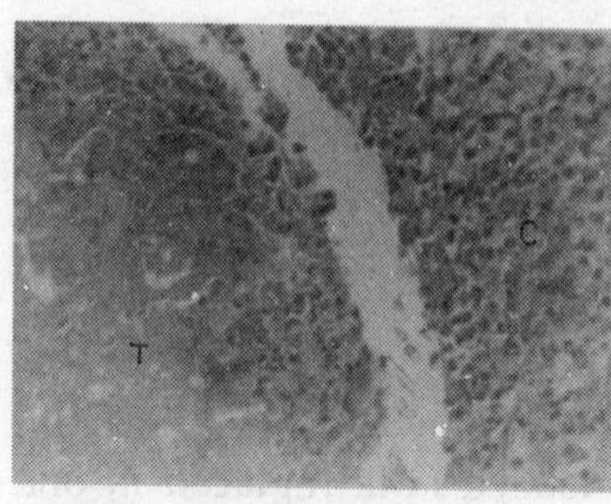

Fig. 15 Valvular thrombus (T)in cattle (vein) separated by a gap from the cardiac musculature (C). Note the homogenous material and different cells in the thrombotic mass. H & E x 400

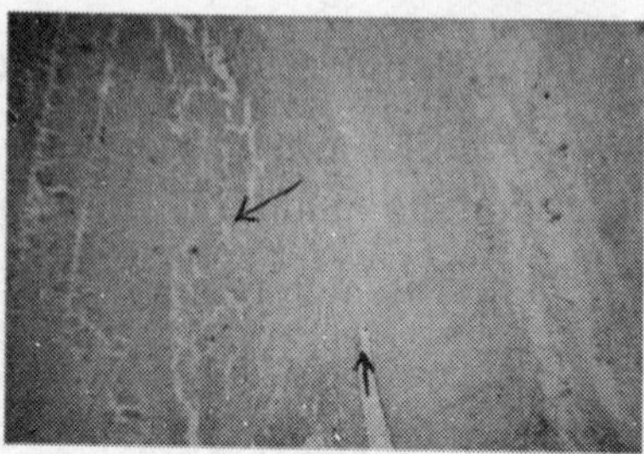

Fig. 16 Thrombosed blood vessel with line of attachment between hyalinised thrombus (at the left) and wall of the vein (at the right) in the udder of a buffalo. H & E x 400.

Differences between antemortem clots (thrombi) and post-mortem clots are as follows :

Antemortem clot	Postmortem clot
1. Formed during life	1. Formed after death usually during the first half hour of death intravascularly
2. Friable, dull and irregularly roughened or with some-what stringy surface	2. Dark coloured, smooth, soft and homogenous like jelly with uniform texture to the naked eye. Also called currant jelly clot.
3. The clot colour is a mixture of red and grey and a clot consists of layers or laminations	3. Not laminated or not adhering to the lining wall of heart and blood vessels etc. (i.e. easily freed from its position)
4. Attached to the vessel wall (i.e. intima)	4. Chicken fat clot is formed slowly with pale portion due to settlement of the formed elements (white cells). Such clots are formed in cardiac chambers. Chicken fat clot has shiny surface of the currant jelly due to heavy iron containing red cells and has also a portion with yellow colour of plasma. In yellow part, mostly light leucocytes are found.
5. Microscopically, the thrombus has blood platelets accumulated at the site of the thrombus formation. Platelets adhere to each other and to the vessel wall. An amorphous pink or grey staining mass which tends to alternate irregularly with layers of red or white cells.The clot is irregularly laminated and an area of attachment can also be seen. The fibrils are thick or dirty pink in the thrombi.	5. Coagulation process is the same as seen in thrombosis
	6. Postmortem autolysis of endothelial cell, Red cells release tissue thromboplastin which is essential for clotting.
	7. Microscopically, it consists of red cells, leucocytes fibrin etc.There is a uniform mixture of these elements in the clots. The heavy red cells settle due to gravity in the lower part ofthe vessels or cardiac cavities during clotting. This process is noticed in the formation of currant jelly clot.

Types of thrombi

Thrombi can be divided into the following types :

1. Red thrombi (Currant jelly thrombi).

Theses thrombi consist of red cells, white cells, fibrin and look like postmortem clots. These are formed rapidly and are called hyaline thrombi after loss of the blood pigments. Red thrombi are formed rapidly in the slowly flowing blood whereas white thrombi are formed slowly in the rapidly flowing blood in the vessels.

2. White thrombi (Pale thrombi)

These are usually white, or with pinkish tint and may be opaque, structureless or have consistency of rubber due to slow formation. Fused platelets, homogenous or hyalinised areas fibrin and leucocytes are present in such thrombi and are found attached to the vascular wall by connective tissue proliferation. (Fig. 15 - 16.)

3. Mixed thrombi or Laminated thrombi

In these thrombi, there is a mixture of red and white layers (alternately arranged in such clots). These are found in aneurysms and look like sliced onion.

Microscopically, pale thrombi consist of fused platelets and leucocytes can be found in the clefts or spaces within such thrombi. When the blood flow is fast, a pale thrombus is formed.

The nidus or nucleus of such thrombus is formed due to conglutination of the platelets. It increases in size due to further accretion or addition of the platelets etc. Heart vegetatious are the examples of pale or dense thrombi.

Microscopically, the red thrombi consist of meshes or fingers formed by trabeculae leading out from the thrombotic mass in which leucocytes, fibrin etc. accumulate with further coagulation, pure red thrombus is an example of simple coagulation.

Thus, it can be seen that platelets have important role in thrombosis due to their properties of adhesiveness.

4. Occluding thrombi

These frequently occlude the lumina of the affected vessels and may be canalized due to openings in them through which the blood passes ahead in the channel. These canals are lined by endothelium which is continuous with that of the vessel on either side of the clot and thus, the circulation is restored (Fig. 17).

5. Obturating thrombi

These thrombi are found distal to their areas of attachment and not attached any more to the wall with their free ends trailing downstream with the blood current in a vessel.

Causes

These are as follows :

1. Injury to the endothelium (lining) of the blood vessels. (a concept of the Virchow's triad)

Injury to the lining of the blood vessels arises from factors like bacteria, toxins, chemicals etc. circulating in the blood stream. Disease in the vessel walls, e.g. arteriosclerosis, parastic aneurysms (i.e. larval infection of *Strongilus vulgaris*) or inflammatory conditions of the blood vessels favour clot formation. Pressure from the outside as in the case of tumours and ligatures etc. give rise to thrombosis.

2. Roughness of the vessel lining

It favours lodgement of the platelets which release thromboplastin on disintegration. The thromboplastin so formed initiates the clot formation.

3. Slowing or stasis of the blood flow

A thrombosis forms with difficulty in fastly moving blood but in condition of stasis or chronic passive congestion as seen in cardiac inefficiency, thrombosis can develop. In inflammatory conditions, there is decelerated blood flow or stasis and localized venous hyperaemia and injury to the vascular endothelium is also caused. These conditions help thrombosis. Such conditions are found in cases of the uterine vessels in metritis. Slow death is followed by formation of clot called agony clot in auricles (e.g. right auricle) or right side of heart which considered to be an intermediate type between antemortem and post mortem clots.

5. Disruption to the laminar flow of blood

This can act as a predisposing cause. Normally, platelets and blood cells are separated from the vascular wall by a zone of plasma but this arrangement is disturbed in case of aneurysms or lodged emboli in the vessels. Blood cells come in contact with vascular endothelium and adhere to it and thus, a predisposing state is provided for clotting.

6. Changes in the composition of blood

These predispose to clotting. Increase in number of platelets or its adhesiveness after surgical operations, parturition and accidental trauma etc. enhance chances of thrombosis. In polycythaemia with an increased viscosity of blood, there is a strong tendency of thrombosis in human beings.

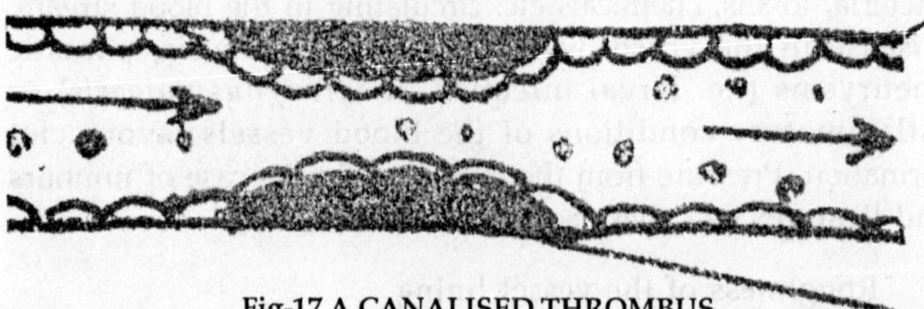

Fig-17 A CANALISED THROMBUS

Effects

1. Ischaemia (Gr. = to keep back blood) from lodged thrombi occurs in the related drained area. Later, infarction follows.
2. Coagulation necrosis (infarction) due to complete occlusion of an artery.
3. Proliferative areteritis, thrombosis arising from lodged thrombi and aneurysm in the blood vessels like anterior mesenteric artery.
4. Thrombosis of iliac arteries in hind-legs in horses with onset of lameness in horses.
5. Oedema in the part drained by thrombosed veins.

6. Canals or channels are formed in the thrombi to establish circulation and these are lined by endothelial cells. (Fig.17)
7. Formation of phleboliths and arterioliths. The thrombi may be calcified with calcium salts in veins and arteries to form phleboliths and arterioliths respectively.
8. Emboli are produced due to fragmentation of the intra-vascular clot. The clots break into pieces which are carried away in the circulation as emboli and are stopped at some bifurcations not permitting further movement and degenerative changes occur due to venous thrombi being lodged in small branches of the pulmonary artery in the lungs.

Organisation of thrombi

Thrombi may be liquefied by enzymes and finally disappear. They may be even hyalinised or organized into fibrous connective tissue. Thrombi are attacked by growing connective tissue cells, capillaries from the vasavasorum and lining endothelium. Thus, they get organized due to growth of the granulation tissues. The thrombi are, later, replaced by fibrous tissue called scar.

To sum up, thrombosis (Gr. = plug) is a physiologic opposite of haemostasis which is involved in maintenance of blood in a fluid (clot free condition) in the blood vessels. It induces a quick localized plug at the damaged site of vascular injury to stop further loss of blood. Injury to tissues is followed by an initial constriction of blood vessels but the activation of platelets and coagulation system causes formation of a haemostatic plug (a laudable process). Platelets become activated at the site of vascular injury and adhere to subendothelial matrix and secret granules to form a plug. Clotting system is activated by certain factors e.g. a tissue factor secreted by endothelium and platelet factor released from platelets. The activated clotting system activates thrombin which converts fibrinogen into an insoluble product called fibrin forming local deposit. Platelets and fibrin aggregate to form a solid plug to close the cut (rhexis) in the blood vessels to check further loss of blood.

The three important factors of Virchow's triad involved in thrombosis are as follows:

1. Endothelial injury.
2. Stasis of blood flow.
3. Blood hypercoagulability is seen due to genetic or acquired disorders, that is, changes in the constituents of blood. A thrombus can be noticed anywhere in the blood circulation e.g. heart chambers, valve cusps, arteries, veins and capillaries. Arterial thrombi grow in a retrograde direction from the point of initial thrombotic attachment. But the venous thrombi propgate in the direction of blood flow e.g. towards the heart from the point of thrombotic attachment. Thrombi reveal lines or striae of Zahn i.e. alternative pale layer of platelets admixed with some fibrin and darker layers composed of red cells. This line indicates thrombosis at the site of blood flow with its base of attachment to the vessel wall.

EMBOLISM

(from the Greek word embolus-wedge)

It refers to an occlusion of blood and lymphatic vessels with bits of foreign substances floating in blood or lymph. These bits of transported foreign materials are called emboli. Fat droplets, gas bubbles, fragments of tumour or thrombi are examples of the emboli which can enter into the circulation. Emboli have more serious effects than the thrombi from which they arise. Pieces of thrombi as emboli resemble thrombi of orgin. These are found lodged at the bifurcation of an artery or in blood vessels which are too small to permit their further escape. Fibrinous emboli, have some amount of fibrin and the process of thrombosis starts at the points of lodgements.

Kinds of emboli are as follows :

Fat emboli

1. These are formed due to sudden release of fat from adipose tissue cells or from fractures of bones or from traumatic wounds

in the subcutaneous fat. In deficiency hepatitis, there is a deficiency of choline, methionine or cystine and the hepatic cells in liver in such deficiencies discharge fat on being ruptured and such fat emboli have been reported in organs like heart, lungs or kidneys, Extensive fatty embolism ends fatally.

2. Gas emboli or Air emboli

When there is a sudden lowering of the ambient pressure, gas emboli are formed. These occur in aviators or divers who suddenly leave their under water compartments i.e. caissons. Large numbers of air bubbles are formed in the circulations which are capable of stopping the blood supply in small capillaries. This condition may end fatally. Air may be sucked into veins through large gaping wounds or may be injected accidentally or willfully into veins. The air emboli in circulation kill the individuals. 10 ml. of air injected in the ear vein of a rabbit kills it almost instantly.

3. Bacterial emboli

These include clumps of bacteria which form metastases or abscesses in the organs. These organisms are stopped in capillaries and may come from heart or septic endometrium and grow at the sites where they are lodged. Such a transfer of infection at a secondary site is called metastasis. This condition is frequently found in liver, kidneys and lungs etc.

4. Detached portions of thrombi, heart vegetations or atheromatous patches act like emboli.

5. Parasitic emboli

Parasites or fragments of parasites act like emboli. Such emboli produce impaction at any site or anywhere in the systemic circulation not permitting their further escape.

6. Tissue cells

Tissue cells from a malignant tumour enter into the circulation as emboli and multiply at the secondary sites in different organs forming metastases there.

7. Spodogenous emboli

These are clumps of agglutinated blood cells as seen after injection of incompatable types of blood in patients.

Impaction sites of emboli

Emboli originating from heart and larger arteries are lodged in systemic circulation. These are found in renal, splenic or cerebral vessels and emboli formed in venous circulation, lymphatic system or right sides of the heart are usually lodged in the lungs. Retrograde emboli are formed by the pulse waves in a direction opposite to that of blood circulation and can be found in extreme venous congestion. Paradoxical emboli are those emboli which pass through patent foramen ovale and, thus, escape from the venous circulation into the systemic circulation without passing through the lungs i.e. such emboli produce impaction anywhere (e.g. kidney) in the systemic circulation.

Effects

(i) Stoppage of the blood supply to an organ. When the lumen of a vessel is blocked due to emboli, ischemia or local anemia, coagulation necrosis and infarction occur in such affected organs. The organs may undergo atrophy. The degree of the effects depends on the richness of collateral circulation.

(ii) Septic or nonseptic bland emboli

In case of pyogenic organisms, abscesses develop at the point of obstruction. Formation of numerous abscesses gives rise to pyaemia. Bland or non-septic emboli are those which do not contain pyogenic organisms. In chronic swine erysipelas, nonseptic emboli are formed.

Fragments of adult canine worms (*Dirofilaria immitis*), and clumps of blood flukes (shistosomes) form emboli in the blood circulation.

INFARCTION

Infarction (L = to stuff in) refers to formation of an infarct and an infarct is an area of coagulation necrosis caused by ischaemia.

Complete occlusion of an end artery in an organ is very serious condition resulting in the death of the affected tissue (i.e. coagulation necrosis). The term end artery means such an artery which has minimum collateral blood supply, e.g. the splenic artery. Presence of collateral blood supply minimizes the effects of the occlusion of an end artery in an organ.

Causes

Embolism and thrombosis are the main causes of infarction. Arterial occlusion, pressure due to tumours on the vessels and ligatures of arteries etc. produce infarcts in the affected tissues due to stoppage of the blood supply. Drugs like ergot may cause narrowing of the blood vessel which leads to formation of an infarct. Infarcts are found in the organs like kidneys, spleen, heart, brain, intestines, lungs and livers etc.

Kinds

Two types of infarcts are found :

(i) Pale, white or anaemic infarct.

(ii) Red or hemorrhagic infarct.

The infarcts may be a few millimeters to several centimeters in diameter. They are roughly triangular or wedge shaped in outlines with the bases directed peripherally. Initially, all infarcts are red or haemorrhagic. Infarcts in the lungs and livers have abundant amount of blood due to dual blood supply in these organs. Infarcts in the intestines are also rich in blood due to good collateral circulation. But in organs like spleen, kidneys and heart, there is a scanty amount of blood supply and poor collateral circulation. Infarcts in these organs blanch within 24 to 48 hours and become pale, white or anaemic. The affected tissues in these organs are swollen and protrude above the adjacent normal tissue due to congestion and oedema etc.

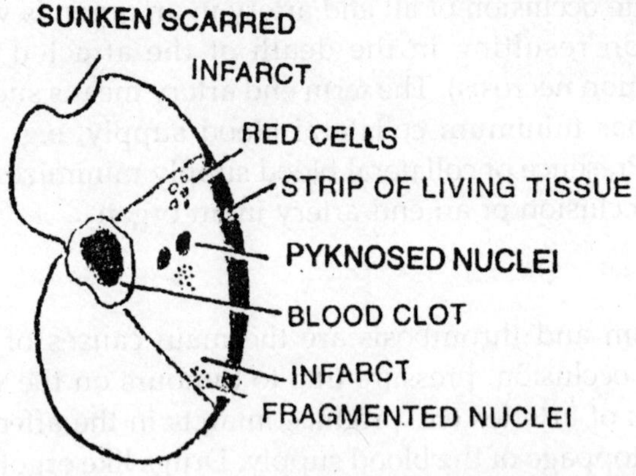

Fig. 18 CONICAL INFARCT IN KIDNEY

Infarcts take a wedge or cone shaped appearance with the apex being at the point of an occlusion and the base being directed towards the surface of the organ (Fig. 18). The outlines of the affected tissues may be irregular due to fusion of adjacent lesions or due to a triangular portion of the organ supplied by the obstructed blood vessel. Microscopically, degenerative changes and necrosis occur in the areas of infarction. Blood diffuses throughout the dead tissue due to escape of blood from the degenerated vessels, giving reddish or purple appearance. The necrotic tissue is infiltrated with fluid from the surrounding tissues and the fluid coagulates with plasma released from destroyed and damaged cells to produce the condition of coagulation necrosis. A zone of congestion and haemorrhage develops immediately surrounding the area of an irritating infarction. Leucocytes emigrate from the neighboring blood vessels and appear in this zone to remove the dead tissue. The initial red colour in an infarct is soon lost due to removal of pigments by leucocytes and the red infarct turns into a pale infarct. A recent infarct tends to fill with blood diffusion from the living areas into the dead capillaries of the infarct. Blood in the efferent veins flows back into the dead necrotic areas to add redness to the colour of recently formed lesions.

Changes in infarcts of different organs

Kidneys

Infarcts in the kidneys are pale or anemic and usually located in the cortex with their bases being near to the capsule. A narrow zone of healthy tissue separates the infarct from the capsule (Fig. 18) and is nourished by the vessels arising from the renal capsule. Infarcts may heal leaving behind a depressed fibrous scar on the renal surface.

Spleen

Infarcts in spleen are almost always haemorrhagic, shallow and sub-capsular. In hog cholera, arteries become occluded due to swelling and hyperplasia of the endothelial cells of the vessels.

Heart

Infarcts in the heart are rare in the animals and more irregular in shape. These may be red or grey. These are not so frequently observed in animals as noticed in humans.

Brain

The infarcts in the brain are usually rare and anaemic and result in liquefaction necrosis and are represented by sharp gap or cavity in the soft tissue.

Lungs

Infarcts in the lungs are not common and result from thrombosis of a branch of the pulmonary artery due to emboli from right side of the heart of venous circulation. Pulmonary infarcts are found in contagious bovine pleura-pneumonia. These are red, firm and standout from the adjacent normal tissues. Necrosis in the pulmonary infarct is absent due to the collateral circulation. Microscopically, alveoli are filled with blood. Emboli from heart valves may produce infarction if congestion to draining pulmonary veins is also present due to cardiac diseases like mitral stenosis. It has been found in cardiac diseases causing congestion in the lungs in human beings.

Intestines

Infarcts in the intestines are always haemorrhagic and large amounts of blood diffuse into the intestinal lumen through the dead tissue. Such infarcts arise from strangulation of the bowel due to hernia or twist. Intestinal infarction leads to fatal gangrene.

Livers

True infarcts in the livers are almost non-existent due to dual blood supply by portal vein and hepatic artery. Pseudoinfarcts develop usually as areas of venous and sinusoidal engorgements on a multilobular scale. Affected tissues lack the features of necrosis.

Summarsing, an infarct which denotes an area of ischemic or coagulation necrosis as a result of an occlusion of either the arterial supply or the venous drainage in a particular tissue or part of an organ is classified into red and white infarcts. Infarcts may be either septic or bland. Tissue infarction e.g. pulmonary infarction is a serious disease in man. Renal cases of infarcts are reported in animals like dogs. Cardiac infarct is rarely noticed in animal. Infarcts are mostly caused by thrombotic or embolic events and arterial occlusion as the most important cause of infarction (a process causing infarcts in organ). Loose tissues e.g. lungs or tissues with rich collateral circulation show development of red infarcts. White infarcts arise from arterial obstruction in solid organ like heart, spleen and kidneys etc. Solidity of these organs limit the amount or extent of hemorrhage in such organs. An infarct is red in colour because of seepage of blood from the adjacent capillaries into the zone of coagulative necrosis. All infarcts tend to form a wedge shaped structure of which apex is formed by the occluded blood vessels and periphery of the affected organ forms its base.

The histologic features of infarcts are similar to those in an ischemic coagulative necrosis. An inflammatory response is noticed around the margins of infarcts which act like inciting irritants.

Reparative processes are noticed in infarcts. Neutrophiles and macrophages remove the dead or necrotic debris. And the process of fibrosis replaces necrotic tissues of infarcts. Ultimately one notices only scars at the site of infarction. Infected emboli from organs like heart in the cases of valvular diseases give rise to septic infarcts.

The vascular supply in an organ, rate of formation of an occlusion, vulnerability of the affected tissue and the blood oxygen content determine the effects of infarcts. Infarcts cause paralytic strokes in humans. Generally, infarcts in organ undergo organization and reparative processes. Only scars are seen at the site of infarcts. Infarcts in intestine, and lungs are always haemorrhagic. Intestinal infarcts are causes of fatal gangrene in animals e.g. horses.

The fate of infarcts

Changes in infarcts are as under :

1. Suppuration and abscess formation in the dead tissue due to pyogenic bacteria.
2. Removal of the infarct by organization and encapsulation around the dead tissue.
3. Autolytic softening of the infarct in brain and heart produces the condition of encehalomalacia and mylomalacia cordis respectively.

Chapter **5**

Constitution and Others

CONSTITUTION

It refers to a state of resistance developed from a complex action of the external causes of disease with diverse internal factors acting through hereditary and environmental influences (Rubarth, 1964). One can consider it as a balance between the weakness and strength of the body. Genetic characteristics determine the individual physiological characteristics i.e. constitution of a particular individual. Endocrine secretions (hormones) influence constitution to a great extent. Constitution comprises of the aggregate of an individual heritable physiological characteristics. It imparts either predisposition for or resistance against a disease to an individual. The resistance of an individual prevents diseases from developing regardless of the environment. Intrinsic factors (predisposing factors), resistance and environment greatly influence the constitution of an individual. Constitution plays an important role in the aetiology and pathogenesis of diseases and is a complex phenomenon in organisms arising from interaction of the inherited properties in the body with the conditions of its existence. Nervous system also seems to influence constitution in the pathogenesis of diseases in the body.

In short, the main features or concepts of body constitution are as follows:

1. An aggregate of stable and heritable property.
2. No change in the constitution all through an organism's life. But there can be variation in the constitution with the condition of an organism's life.
3. Fixation of some properties in the progeny of an organism due to prolong influence of extrinsic factors.

4. Adjustablity of an organism to the changing condition of its environment.

FEVER (FEBRIS)

Fever is a general reaction of warm blooded animals and man to the action of harmful and infectious agents. In this, there is a disturbance in the heat regulation with elevation of the body temperature irrespective of the temperature of the external environment. Hyperthermia occurs under the influence of elevated temperature of the hot or humid external environment but fever in an organism is seen under usual temperature conditions. There is a disturbance in central thermo-regulation leading to accumulation of heat and elevation of the body temperature. In an elevation of body temperature, there is a temporary predominance of heat production over heat loss. In short, fever is a disorder in metabolism marked usually by pyrexia and functional disturbances. Fever is noticed in very old age without rise in the body temperature. It may be considered as an organism's struggle against an infectious agent. Pyrexia refers to the elevation of temperature as seen in fever or hypethermia in hot days of summer session. Some cases of fevers may not show pyrexia.

Hyperthermia

It means an elevation of temperature which is not associated with fever. It is seen in cases of heat stroke and certain nervous lesions and in conditions of highly elevated external temperature. Animals transported by railway or truck may show hyperthermia. Belladona and cocaine produce hyperthermia. Hypothalmic lesions destroy or disturb the temperature regulation centre. Hyperthyroidism or an excessive exercise can produce fatal hyperthermia.

Hypothermia

In this, there is a sub-normal temperature and is seen in cases of starvation, anemia, shock and collapse. It may arise from cold climate (marked by a lower temperature in the atmosphere).

Causes

Toxins (bacterial), protozoan or viral infection are chief causes of fever. Fever may also be caused by proteins formed in tissue necrosis or fractures of bones. Haemolysis, salts, drugs and neurogenic factors influence the processes of thermoregulation in the body.

Stages of fever

Three distinct stages in fever are as follows :

(i) The initial cold or shivering stage. There is a feeling of cold and rigors due to contraction of surface blood vessels in the skin. In shivering stage, temperature starts rising. One also notices cooling of the superficial layer of the skin during chills.

(ii) Hot stage

In this stage, the body temperature reaches at its maximum. The blood vessels in the skin dilate and the patients feel heat.

(iii) Sweating stage

It is characterized by a fall in temperature. The temperature falls abruptly in few hours in crisis but in lysis, it falls gradually in several days.

PATHOLOGIC CHANGES DURING FEVER

Circulation

There is an increase in pulse rate due to stimulation of the sympathetic nervous system. In the first stage, the blood pressure is slightly elevated but it is lowered in the later stages. There is usually an increase of 8-10 beats in the heart rate for a 1^0 rise in temperature in organisms e.g. as in humans. But there are also reverse phenomena.

Respiration

It is marked by an accelerated respiration. Rise in the level of carbondioxide in blood is due to increased metabolism. Increase in respiration helps in physical thermoregulation of fever.

Secretions

There is a fall in secretions of skin, liver, pancreas and salivary glands etc. Urine is secreted in decreased quantity and its specific gravity is increased. There is an increased secretion of ammonia, urea, phosphates and sulphates and albumin and ketone bodies may appear in urine. One notices retention of water by the tissues.

Tissue changes

A marked wasting of fat with muscle atrophy in body occurs. Due to toxins, dystrophic changes like cellular swelling and fatty changes are seen in organs like kidneys, liver and heart etc.

Heat stroke

Under various situations, animals fail to eliminate sufficient heat to maintain body temperature. Unusual heat and humidity expose the animals to heat stroke as seen in shipping or transportation in a small poorly ventilated crate or confinement of several animals and humans in truck or bus etc. Such conditions may subject the animals to high temperature and increased humidity. Hot and humid weather may cause heat retention or heat stroke in animals.

Affected animals show high temperature (106°C to 108°C F). Respiratory distress, discomfort, congested mucosae, tachycardia and excitement. Collapse and death are seen in cases of heat stroke.

Gross changes

Lungs are oedematous with focal consolidations of bronchopneumonia. Heart, kidneys, meninges, lymph nodes and muscles are severely congested. Extensive and rapid autolysis takes place in the animals after their death depending on the external environment.

Microscopic changes

Congestion and oedema are seen in the lungs. Livers are congested with dissociated hepatocytes and centrilobular necrosis. There is presence of subendocardial haemorrhages. Capillary congestion is found in brain and kidneys etc.

To sum up, fever is a complex response as a sign or symptom of several infectious diseases or inflammatory conditions. It is marked by a disturbance in the temperature or heat regulating system located in the preoptic anterior hypothalamus. Antipyretic drugs reduce body temperature in fever but these drugs have no effect on hyperthermia (a condition marked by high elevated body temperature).

Some important features of fever are :

1. Inflammatory cytokines acting on preoptic anterior hypothalamus causes fever. Chill is a feature in febrile conditions. The cooling of the superficial layer of skin produces shivering or chill .

2. Reduction of body temperature in fever by antipyretic drugs

Fever is noticed in organisms like men, animals, fishes, and amphibians, reptiles, birds and several other animals. Mammals in fever respond to antipyretic drugs. In cases of infections, newborn animals may not show fever. Bacterial endotoxins and cytokines like IL-1 and TNF-α induce fever in animals. Fever seems to influence or reduce bacterial activity in infected animals and by doing so, it has a laudable influence on the body.

In short, infectious fevers are caused by bacteria, toxins, waste products and inflammatory reactions. Pyrogenic substances in bacteria produce fever in the body Even extracts of neutrophiles may cause fever in the body but no such effect is shown by the extract of macrophages. High body temperature or a sudden drop in the body temperature is very harmful to a patient.

Hyperthermia (overheating) in the body arises from a disturbance in thermoregulation and impediment to heat dissipation in warm environment. Hot and humid climate limits heat loss and causes retention of heat in the body producing a condition called hyperthermia. A hot day with humid air causes hyperthermia or heat strok in animals. Petechial haemorrhages in the meninges and several tissues are noticed in the fatal cases of heat stroke .

Effect

Death may happen. Increased phagocytosis and rise in production of immune bodies are also noticed.

Disturbances in water metabolism

Consumption of water by an organism, formation of water in the process of metabolism, ability of the tissues to retain water and excrete it through the kidney, skin, lungs and intestines are determinants of water balance in the body. A positive water balance means retention of water in the organism.

The main causes of disturbances in water metabolism are :

1. Excretory dysfunction of the kidneys and skin.
2. Disturbances in the inter-change of water between the blood and the tissues.
3. Hypofunction of the thyroid.
4. Increased production of adrenocortical hormone and mineralocortiocoids. Mineralocortcoids stimulate re-absorption of sodium and depress reabsorption of potassium facilating the retention of water.
5. Food rich in carbohydrates and proteins.
6. Intake of excessive amounts of sodium chloride.

Negative water balance is the cause of dehydration of an organism. Insufficient water consumption, excretion of excess of water, (for example, profuse perspiration, copious urination, frequent diarrhea or recurring vomiting) are causes of negative water balance. Dehydration is recognized by loss of weight, thirst, haemoconcentration, increase in the dry residue, and specific gravity of the blood, relative increase in haemoglobin, diminished oxidative processes, reduced blood pressure and intoxication etc. The tissues loose their turgor. The mucous membrandes are dry and the eyes sink in and the activity of the central serous system in disturbed.

The loss of 20% of water in body is dangerous to life. The diencephalon and hypophysis regulate water metabolism in the tissue. And-diuretic hormone secreted by the posterior lobe of the hypophysis establishes the level of water retention by the tissue, water re-absorption by the tissues and water reabsorption in the renal tubules. Diabetes insipidus caused by an injury to hypopysis or diencephalon results in copious urination and intense thirst.

STASIS

In Greek, the word stasis means standing and the stasis is a complete cessation of the blood flow. The vessels are dilated and filled with closely adhering red cells.

Stasis is of two kinds :

(1) Venous stasis

(2) Capillary stasis

1. Venous stasis

There is an impeded outflow of blood through draining veins. Blood flow is slowed down and vasomotor-nerves are paralysed.

2. Capillary stasis

It can occur irrespective of any obstruction to the outflow in a vessel. It is caused by tissue desiccation (exposed peritoneum), heat or cold, acids, alkalies and mustard or croton oil. Infection can cause capillary stasis. Vasomotor, chemical and physical disturbances cause capillary stasis. Excessive production of histamine occurs due to action of harmful agents on the tissues and causes capillary dilation favouring capillary stasis or stoppage of the blood flow.

The results of stasis

1. Blood flow may be restored after climination of the cause of stasis.
2. Blood stasis may be irreversible and the corresponding part may be necrosed.

Shock (Cardiovascular collapse)

The term shock (middle Eng. and Fr. = to shake or jolt) is defined as an acute peripheral circulatory failure. Shock is an entity or state with multiple causes. There is an inequality between blood volume and the capacity of the vascular system to retain blood in shock. In shock, the volume of circulating blood is not adequate to fill the space in the vascular system and the heart does not get sufficient blood to fill its cavities for normal cardiac activities. This happens in case of haemorrhage or in burns due to loss of copious serous exudate. The excessive loss of fluid or blood may happen in case of operation. Injuries and considerable exposure in handling of intestines often leads to the fatal state of shock.

CLINICAL SIGNS

The clinical state of shock refers to the presence of the changes like marked functional depression, pallor, sweating, feeling of cold, lack of strength and sub-normal body temperature.

There is no warmth on the body surface with fall in blood pressure. Shallow, irregular respiration, rapid, weak and irregular pulse of failing heart are noticed.

In such cases, death may follow any moment. Histamine released into circulation from extensive traumatic and an inflammatory changes in tissue causes arteriolar contraction and dilatation of the capillaries. Each expanded capillary holds more blood (i.e. more red cells etc.) than normal. The dilation of the capillaries is seen in the internal organs (mainly in the abdominal region) and this state leads to a drop in blood pressure. Destruction of animals tissues, for example, the hind limbs of an animal, can lead to death within less than an hour.

Shock can be divided into the following types :

1. Primary shock or neurogenic shock
2. Haematogenous or secondary shock

Primary shock

It occurs in immediate conjunction to trauma. The characteristic sign is unconsciousness. It arises from voasodilatory effects of histamine. Pain can also produce it.

Secondary shock

It can occur in 2 to 24 hours after trauma and results from haemoconcentration due to loss of plasma through vascular walls. Blood volume gets reduced but there is a consequent rise in fluid in the tissues and cells. The vasomotor centre gets paralysed and blood pressure drops futher. Hypoxaemia and damage to the capillary walls arise from inadequate circulation of blood.

Postmortem lesions

These are as follows:

1. Severe congestion of the vessels in the lungs, liver and intestines
2. Ischaemic state (pallor) of the peripheral parts of the body
3. Pulmonary oedema (oedema fluid and haemorrhage in the alveoli)
4. Oedema and haemorrhage in the intestines
5. Some bloody fluid in the peritoneal, pleural or pericardial cavities
6. Empty and bloodless spleen
7. Degenerative changes in the kidneys, heart, adrenal glands, lymph nodes and spleen
8. All parts in the body are ischemic with little signs of congestion in cases of shock from severe hemorrhages
9. Considerable serum exudation in the bruised tissues.

Summarizing, extensive trauma, severe haemorrage, bacterial sepsis, myocardial infarction and burns etc. are important causes of shock in man and animals.

Main features of shock (a multiple organ systems failure) are as follows.

1. Tissues hypoperfusion, hypotension and cellular anoxia.
2. Low cardiac out put is noticed in cases of myocardial pump failure, and hypovolemic shock.
3. Cardiac tamponade (extensive compression on the heart noticed in haemorrhages and traumatic pericarditis).
4. Loss of blood plasma causing hypovolemic shock.
5. Systemic bacterial infections cause septic shock (endotoxic shock) due to gram negative bacterial or fungal infections. Endotoxins (e.g. bacterial wall polysaccharides-LPS) released from degradation of bacterial walls cause septic shock. The main components of LPS consist of lipid A (a toxic fatty acid) and O antigens (a polysaccharide coat). Septic shock can be induced by injections of LPS. A complex of LPS with blood protein binds to CD14 molecules on leucocytes like monocytes and macrophages and endothelial cells etc. LPS- binding protein complex activates cells like leucocytes and vascular wall cells to elaborate different kinds of cytokine mediators leading to development of the pathologic state of shock.

Results

i. Presence of general or myocardial ischemia.

ii. Death may a follow.

The following table gives an information about cytokines generated by leukocytes and endothelial cells in causing state of shock in the body:

TABLE-5 Cytokines generated by leucocytes and vascular endothelial cells in pathogenesis of endotoxic or septic shock and theire effects in bacterial infections.

Cytokines generated / Other factors		Concentration of cytokines	Effects
TNF, IL-1, IL-6,IL-8, No (nitric oxide) and PAF (platelet activting factor) induced by lypopolysaccharide (i.e. a LPS binding protein complex to CD 14 molecules on monocytes).		a. Low quantities of cytokines produced	1.Activation of monocytes monocytes and macrophages phages 2. Activation of vascular endothelium 3. Activation of complements like C3 a and C5a (anaphylatoxins) All the afrosed aforesaid facts mediate local inflammatory response in tissues
Activated cells	Cytokines and other factors produced.		
1. Monocytes (activated by LPS)	TNF and IL 1 (induced by TNF)		
2. Endothelial cells (activated by an IL-1and TNF)	IL-6, IL-8 and endothelial adhesion molecules		
		b. Moderate quantities of cytokines	1. Fever and acute phase reactants released by liver 2. Production of TNF and IL-1
		c. High quantities of cytokines	1. Low cardiac output and low peripher peripheral resistance 2. Injury to blood vessels 3. Thrombosis and pulmonary pulmona distress. 4. Septic shock as seen in peritonoitis, endometritis and absc an abscesses

Results

1. Presence of cerebral or myocardial ischemia.
2. Death may follow.

Chapter 6

Inflammation

Inflammation (L=to set on fire) is a complex tissue reaction of the body of an organism to the action of irritants or noxious agents. The reactions to tissues manifest themselves in functional and structural changes in the vessels and tissues and include three main closely inter-connected and simultaneous phenomena. These are, namely, (1) tissue dystrophy (alteration), (2) circulatory disorders (i.e. exudation and emigration) and (3) proliferation of cells. All these processes are directed towards adjustment, defense and restoration of the functions to normalcy. In short, inflammation is a reaction of tissues to injury. Injury to the tissues is followed by inflammation. Infections (bacterial, toxic and viral etc), mechanical (gun shot wounds), thermal (burn or frost bite) and chemical factors (acids or alkalies) are causes of inflammation. Injurious agents are, thus, living and nonliving factors.

The intensity of an inflammatory reaction depends on the following facts:

1. Reactive properties of the organisms or intensity of the irritants.
2. Anatomic and physiologic characteristics of the affected tissues.
3. Duration of activity of the irritants or noxious agents at the injured sites.

The reactions at the sites of injury tend to destroy or limit the spread of the injurious agents and are also directed towards repair or replacement of the damaged tissues. These local reactions to injury constitute what is known as inflammation. Inflammation can be said to be preceded by an injury and has laudable, beneficial or defensive effects.

Inflammation is a local defensive reaction in a narrow sense and its external signs are redness (rubor), swelling (tumor), heat (color), pain (dolor) and dysfunction (function-laesa). These are characteristics of acute inflammation in the skin and mucous membranes.

The foregoing signs are either feebly marked or not seen in the internal organs like liver, kidneys and heart etc. The external cardinal sign of redness is camouflaged by the organs normal colour. There may be no swelling, redness, heat or pain in chronic inflammation as seen in the case of fibrosis (cirrhosis) of kidneys and liver etc.

The pathogenesis of the cadinal signs of inflammation i.e. rubor and tumor etc. are as follows :-

Rubor

It arises from a greater flow blood through arteries and capillaries. An active hyperaemia is a very important inflammatory reaction.

Tumor

It originates from loss of fluid from the blood vessels in to the surrounding tissues and cellular infiltration and is marked by the development of a swelling.

Color

Its causes are active hyperaemia, local heat producing processes and molecular changes etc. in the tissues.

Dolor

It originates from pressure exerted by the local swelling on the sensory nerve endings in the region and harmful effects of the irritants causing inflammation.

Functio-laesa

It refers to an impaired function arising from tissue damage. The latin suffix "itis" is added to the Greek or Latin appellation of the affected organ or tissue. In designating the various kinds of inflammation as in cases of nephritis (inflammation of the kidneys) and dermatitis (inflammation of the skin) etc. the suffix 'itis' has been used.

Inflammatory processes involve the vasculariesd connective tissue and its surrounding parenchyma of an organ. Fluid components and cells etc. escape out into the tissues of an organ or on its surface and, thus, the volume of the tissue is increased. In old cases of inflammation, tissue proliferation occurs. Local changes of inflammation include increased blood flow, escape of fluid and increase in cells in the affected tissues. The main inflammatory changes are as under :

1. Degenerative changes (alterative).
2. Exudative or infiltrative changes.
3. Proliferative changes.

Degenerative changes.

Theses changes are only part of an inflammatory reaction. In inflammation, cellular swelling, fatty changes and necrosis can be found, but in some cases, these destructive processes are not so prominent histologically. Destructive processes can involve both cells and inter-cellular substances. In parenchymatous organs like kidneys, the exudative and proliferative changes are hardly evident before the predominant destructive changes into renal parenchyma. Alterative inflammation is an acute inflammation based on degenerative changes in parenchymatous organs.

Alterative inflammation (Parenchymatous inflammation)

There is a predominance of dystrophy, necrobiosis and necrosis while exudative and proliferative changes are feebly present. This is found in cases of infections and intoxications in organs like kidneys, liver and heart etc.

Vascular and exudative changes

In some acute inflammations, exudation and infiltration are dominant changes. There is hyperaemia with escape of fluid and cells from the blood vessels into extravascular spaces. The exudate consists of fluid and cells escaped out from blood vessels, lysed cells and intercellular substances and dead bacteria etc. The fluid and cells of an inflammatory reaction present on the surface of an organ or in a serous cavity is called exudate whereas the term infiltration refers to some fluid and cellular components which have escaped from the blood vessels into the interior of tissues.

Acute inflammation can be classified into the following types on the basis of exudates :

1. Serous inflammation
2. Fibrinous inflammation
3. Purulent inflammation
4. Haemorrhagic inflammation
5. Catarrhal or mucous inflammation
6. Lymphocytic inflammation

Inflammation can be classified according to its course or duration of illness into actue, subacute and chronic. Acute inflammation has a brief course and had intense inflammatory reaction and the main changes are degenerative and vasoexudative in nature. Chronic inflammation runs a protracted course with mild symptoms and is characterized by proliferative changes. Subacute inflammation adopts a middle or intermediate course or a course between acute and chronic inflammations. Acute inflammation is all of a sudden in its onset and may last over a few hours. There may be rapid recovery or death within a few days. The cardinal signs of inflammation like redness, swelling, heat and pain etc. are noticed in this type. Chronic inflammation has a longer duration and may show some signs of acute inflammation. Diseases are also divided according to their duration into acute, subacute and chronic. Acute diseases last over a short period from several days to 2-3 weeks and subacute ones from 3 to 6 weeks. Chronic diseases extend over more than 6 weeks, and these frequently arise from acute diseases as sequalae.

Circulatory disturbances

The inflammatory agent acts on the receptors of the affected part and produces reflex circulatory disturbances.

(1) Momentary contraction

At first, there is a brief vascular spasm. Stimulation of the vasoconstrictor nerves gives rise to vascular spasm and pallor of the injured part. It disappears rapidly i. e. it is a transient phenomenon.

Active hyperaemia

The vascular spasm is followed by dilation of arterioles and capillaries. Capillary circulation increases and more blood is brought to the inflamed part to produce arterial active hyperaemia. The increased blood is responsible for greater haemorrhages in injuries to the vessels of the inflamed region. There is a lesser utilization of blood oxygen by the tissues of the inflamed part. Redness of the inflamed part and increase in temperature are due to an increased blood flow to the part. Increased tissue metabolic processes also add to rise in temperature of the affected part. The dilation of the vessels follows their transient constriction and is due to the following factors :

1. Vasoactive amines and reflex effect of the harmful agent. (a triple response of Lewis, 1927)

Sudden dilation of the arterioles and increased flow of blood to the periphery of the inflamed focus is due to action of vasomotor apparatus i.e. role of axon reflexes in vascular dilation. The arterioles outside the injured areas are dilated due to axon reflexes and are responsible for bright red halo around central dull red area which is due to dilation of the minute vessels (capillaries). Capillary dilation is caused by histamine and serotonin (5-hydroxytryptamine) formed in the injured tissues.

2. Rapid paralysis of the vasoconstrictors and excitation of the vasodilators.
3. Increased concentration of the hydrogen ions. The acidity developing in the focus of inflammation causes vasodilatation. Tissue acidity also causes an increased permeability of the capillary walls for protein so that the larger

molecules of proteins pass through the endothelium into the extra vascular tissues.

4. Electrolytic changes.

An increase in potassium ions also causes vasodilation.

5. Metabolites and products of tissue disintegration.

These substances have vasodilatory effects.

Slowing of the blood stream

An increase in the blood flow is, later, followed by slowing down of the blood current. The dilated blood vessels become engorged with blood and a state of congestion develops. The causes of slowing of the blood current are as follows :

1. Partalysis of the neuromuscular apparatus of the vessels.
2. Increased permeability of the vessels and greater transudation of fluid from the vessels into the tissues.
3. Concentration of blood and an increase in its viscosity.
4. Resistance to blood flow due to roughness of the internal wall of small vessels arising from the adhering leucocytes
5. Some swelling of the endothelial cells and formed elements of the blood.
6. Mechanical obstacle to the outflow of blood from compressed small veins due to oedema in the surrounding tissues.
7. Formation of thrombi and occlusion of the vessels.
8. Reduction in the volume of circulating blood in the dilated blood vesels.

Later, stasis (i. e. blood standing) may develop due to slowing of the blood current.

Exudative and emigrative changes

Exudation (a passage of protein containing fluid through the walls of vessels into the extravascular tissues) results from dilatation of the vessels and slowing of the blood current. Exudate is a fluid that passes from blood into the tissue around blood vessels and consists of leucocytes, red cells, platelets and products of tissue disintergration (Fig. 19)

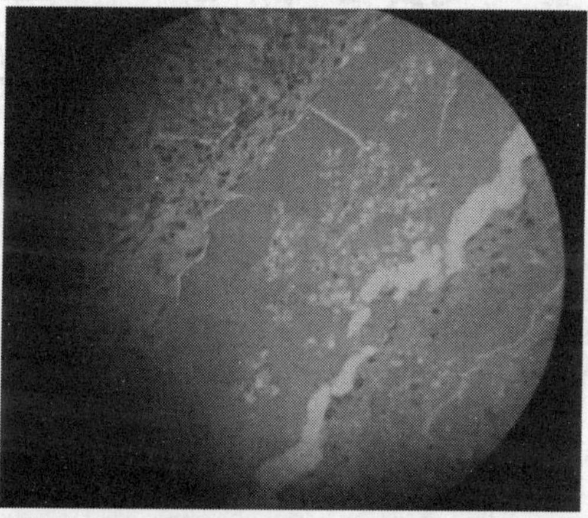

Fig. 19 Oedema of the gall bladder, pig. Note the proteinous exudate containing inflammatory cells (H & E x 400).

The causes of exudation are

1. Increased permeability of the capillaries.
2. Elevated blood pressure in the vessels of the inflamed focus.
3. Increased colloidal osmotic pressure in the inflamed tissue. It increases the escape of fluid from the blood vessels in the surrounding extravascular tissues.

The increased permeability of the capillaries are also due to chemical mediators like amines etc.

Presence of H-substance in the injured tissue and 5-hydroxytryptamine, proteases such as plasmin, kallikerin and globulin permeability factor and polypeptides as leukotaxins, bradykinin and kallidin enhance permeability of the capillaries.

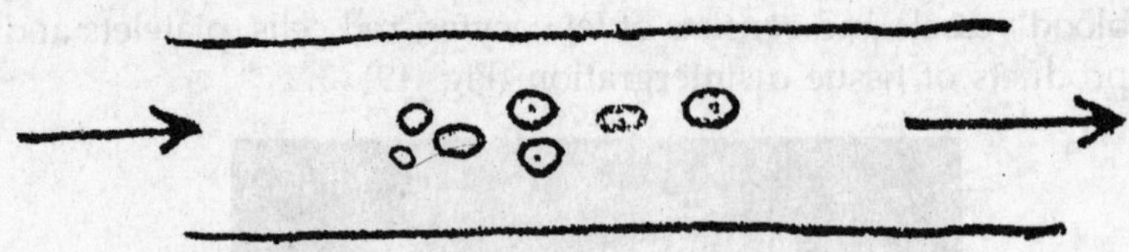

Fig. 20 NORMAL FLOW OF BLOOD WITH NO LEUCOCYTES STICKING IN A VESSEL

Emigration or diapedesis (Gr. = through + to leap) refers to the process of escape of the formed elements of the blood through intact capillaries in the surrounding tissues (Fig. 22). Leucocytes have active diapedesis or escape out but, there is a passive diapedesis shown by the red cells.

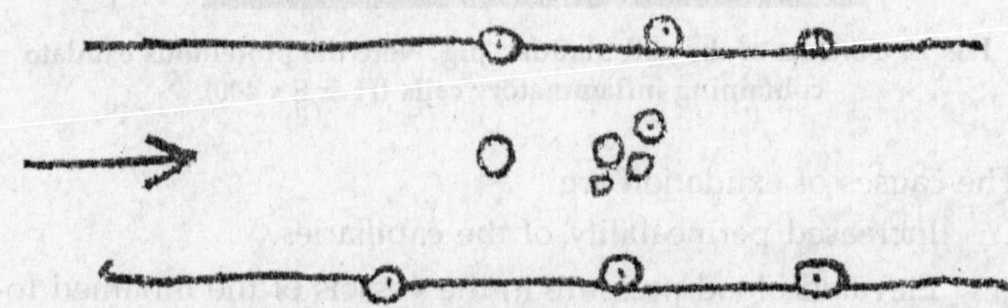

Fig. 21 MARGINATION OF LEUCOCYTES

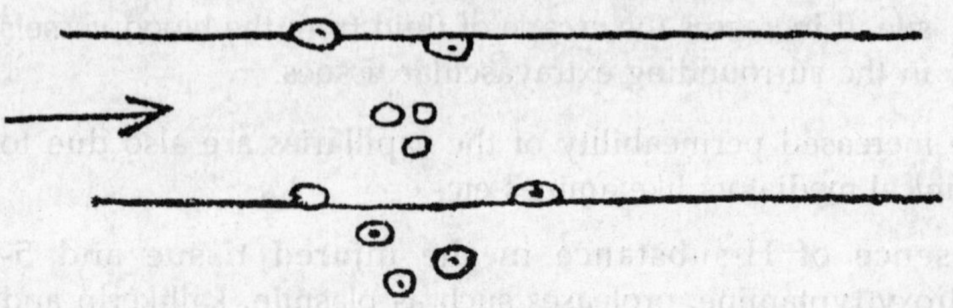

Fig – 22 EMIGRATION OF LEUCOCYTES

When the blood current slows down, the leucocytes tend to accumulate in the plasmatic parietal layer of the vessels in the inflamed focus showing a departure from an axial (central) corpuscular stream (Fig. 20) and these leucocytes tend to adhere to the endothelium of vascular wall (Fig. 21). This phenomenon is called margination of the leucocytes (i.e. due to marginal position of leucocytes). Emigration refers to passage of leucocytes from the vessels into tissue (Fig.22). It occurs simultaneously with exudation. The white cells escape through the walls of capillaries and small veins. The leucocytes previously noticed in the centre of blood stream accumulate in the plasmatic parietal layer and extend narrow protoplasmic protrusions i.e. (pseudopodia) which pierce through the wall of these vessels. The prominences go on to external wall of these vessels. The prominences go on increasing until the leucocytes completely emigrate through the vessel wall and detatch themselves from the vessel walls to enter into the surrounding tissues and can then move ahead towards the focus of inflammation with amoeboid movements. These leucoytes further engulf bacteria, foreign bodies and particles of cell elements (phagocytosis). Some of the leucocytes are destroyed to liberate various enzymes to digest products of decomposition. The surviving leucocytes may proliferate or pass back into blood stream. In serous inflammation, few leucocytes emigrate from the vessels while in the purulent inflammation; the leucocytes emigrate in abundance to engulf bacteria etc. In the beginning of inflammation, neutrophiles (microphases) mostly emigrate and lymphocytes and monocytes (macrophages) emigrate lastly from the vessels. Emigration of leucocytes from the vessels into the tissues is dependent on chemical attraction (i.e. chemotaxis). Bacteria like streptococci and staphylococci and products of disturbed metabolism show chemotaxis which depends on leukotaxine i.e. a substance obtained from the inflammatory exudate. Leukotaxine increases both capillary permeability and chemotactic action.

Leukotaxine is a polypeptide formed due to protein degradation. Certain active globulins possess chemotactic action and/substances like chloroform, benzene and alcohol repel leucocytes. Substances like peptones and organic acids and histamine reduce the surface tension of leucocyte and at the site of the reduced surface tension of the leukocytes; protoplasm begins to protrude or project to form ultimately pseudopodium for the sake of emigration.

The leucocytes which carry negative charges are attracted towards the positively charged hydrogen ions accumulated in the inflamed tissue. Increased exudation also causes emigration of leucocytes.

Adhesion molecules in the vascular endothelium and receptors in leucocytes are very important factors for emigration of leucocytes from the lumens of the vascular channels into the extravascular spaces in inflamed zone of tissues. The main steps in adhesion and transmigration of leucocytes in the inflamed or injured tissues (Table-6) are:

1. Binding of complementary adhesion molecules on the leucocytes and endothelial surface of the vascular channels. It is an interaction between receptors and ligand proteins

Table-6 Interaction between complementary adhesion molecules on the endothelium and leucocytes (e.g. receptor – ligand binding)

Endothelial adhesion molecules	Leuocyte receptors	Effects
1. P. selectin (PADGEM) Location in endothelium and platelets	Sialyl-Lewis X (sialylated form of oligosaccharides	Rolling of leucocytes e.g. neutrophiles, lymphocytes, monocytes on endothelium.
2. E- selecting (ELAM-1) Location in endothelium	Sialyl-Lewis X (sialylated form of oligosaccharides, ESL-1 and PSGL-1 (mucin like glycoproteins)	Rolling of leucocytes and adhesion of neutrophiles and monocytes etc. to activated endothelium
3. ICAM-1 (Intracellular adhesion molecule-1) Location- Activated endothelium by cytokines e.g. IL-1 and TNF-α as a source of its generation and an adhesion molecule of the immunoglobulin family	CD-11/CD-18 (β_2 integrins), LFA-1 (Lymphocytes function associated antigen 1)	Adhesion, arrest and transmigration of all leucocytes

4. VCAM-1 vascular cell adhesion molecule) Location-Activated endothelium by cytokines e.g. IL-1 and TNF-α as a source of its generation and an adhesion molecule of the immunoglobulin family	VLA-4 (β1 integrins)	Adhesion of leucocytes like monocytes, lymphocytes and eosinophiles
5. Glycam- 1 (mucin like glycol protein) CD-34	L-selectin (present in leuocycytes)	Rolling of neutrophiles and monocytes

The main beneficial functions of the exudate in inflamed zone are as follows :

1. Dilution of the concentration of the poisonous substances or toxins in the tissues.
2. Antibodies in the exudate help destruction of bacteria by leucocytes.
3. Formation of mechanical barrier of precipitated fibrin net work. Coagulated lymph prevents microbes and toxins from being absorbed from the inflamed zone.

The harmful roles of the exudate are as under :

1. Excessive absorption of exudates causes rapid spread of bacteria and toxins in the body.
2. Accumulation of the exudate in pericardium, thorax or abdomen may compress the adjacent organs to even fatal extent.
3. Excessive exudate causes swelling of the inflamed part.

The details of different kinds of **acute inflammation** are as under:

1. Serous Inflammation

It is characterized by exudate which is watery and contains a few cells and little or no fibrin. The cells in the exudate include neutrophiles, desquamated or degenerated cells as well as red cells. When the exudate is infiltrated into the tissues, it is called an inflammatory oedema. It may measure in litres in serous cavities. In organs like liver, heart, and kidneys, serous inflammation is characterized by an accumulation of protein rich fluid in the tissue spaces. Serous exudation is due to passage of small protein particles, like albumin and globulin through the capillary walls into extravascular spaces..

Occurrence

1. Serous cavities lined by mesothelium and well vascularised tissues such as peritoneum and thorax.
2. Lungs – Inflammatory oedema is seen in the affected lungs.
3. Skin – Blisters, vesicles, bee stings and burns.
4. Mucous memberane – Vesicular stomatitis (localized serous inflammation).

Causes

1. Trauma, tears, sprains or blows in case of joints
2. Infection in peritoneal and pleural cavities etc.
3. Viral or bacterial infections in the lungs. In lungs, most cases of pneumonia start with exudate formation. *Pasteurella multocida, Brucella bronchiseptica,* viruses of equine and procine influenza may induce such inflammatory changes.
4. Chemical irritants like chloroform.

Inhaled chloroform produces pulmonary oedema.

5. Snake bite- Bite of an American rattle snake produces a marked local inflammatory oedema.

Gross changes

1. Presence of watery fluid in serous cavities or tissue spaces.
2. Small amount of fibrin adhering to the involved surfaces.
3. Red tinged fluid is present due to red cells.
4. Hyperaemia occurs round the zone of inflammation.

Microscopic appearance

These are as follows :

1. A homogeneous pink staining precipitate in the natural spaces.
2. Various kinds of leucocytes and traces of fibrin with the precipitated protein.
3. Presence of hyperaemic or congested vessels in the inflamed zone.

Significance

1. Dilution of toxins or chemical irritants with serous fluid.
2. Presence of excessive serous fluid causes swelling of the part and interference with its functions.
3. Lubrication of the inflamed surfaces or hyperaemic areas or membranes with reduction in pain feeling as seen in painful pleurisy..
4. If the cause is removed, fluid is soon absorbed with return to normalcy.

2. Fibrinous inflammation (Croupous inflammation)

When large protein particles (e.g. fibrinogen) pass out through the capillary walls, the exudate formed is of fibrinous type. Large amounts of fibrinogen clot in the exudate and the fibrin is the most important constituent in such exudate.

Occurrence

1. Found on mucous and serous membranes.
2. Fibrinous exudate is found on the mucous membranes of respiratory system and also in the pericardial sac. The exudate may also be serofibrinous or fibrinopurulent in nature.

Gross appearance

1. A dull and cloudy haze on the serous and mucous surfaces and adhesion of stringy material to these surfaces at a later stage. These shreds may look red due to the presence of blood.
2. Formation of pseudomembrane due to presence of dense and tough layer of fibrin. The sheet looks like a white or yellowish structure.

3. Formation of croupous membrane. This is a layer of fibrinous exudate which can be found in laryngeal infection and the exudate formed is voluminous or suffocating. It is loosely attached to the under lying tissue and the exudation does not extend deep into the tissue and may only extend up to the epithelium.

4. Formation of diphtheritic membrane

There is a firm attachment of the fibrinous exudate to the underlying structures. When it is torn off, a haemorrhagic surface is left behind. In short, a superficial layer of bleeding is visible.

It is found in human diphtheria and calf diphtheria (caused by *Fusiformis necrophorus*).

Microscopic appearance

1. Microscopically, the fibrin is adherent to the underlying serous or mucous surfaces and fibrils can be traced into the epithelial or mesothelial cells of the parent structure.
2. Presence of necrotic tissue, precipitated protein, leucocytes and red cells in the inflamed focus.
3. Hyperaemic underlying tissues.
4. In pulmonary alveoli, there is a formation of a solid non-fibrillar mass and the fibrin stains blue with haematoxylin due to chromatin released from dying or degenerated nuclei. A mixture of serous fluid and fibrinous exudate and fibrinous material and purulent exudate gives rise to serofibrinous and fibrinopurulent exudate respectively in the inflamed tissues.

Causes

1. **Bacteria like** *Fusiformis necrophours and Salmonella* **cholerae suis.**
2. Viruses of avian laryngotracheitis and malignant catarrhal fever.

Effects

1. Fibrin supports leucocytes in the inflamed zone. Fibrin

and dead tissue (underlying) prevents loss of blood and a layer of exudate over the tissue reduces pain in the septic sore throat.

2. Underlying surface is regenerated and the fibrin is dissolved. It may be solughed off as bronchial or intestinal casts.

3. Organisation of fibrin by fibrous tissue can take place.

Adhesions develop in case of pleura, pericardium and peritoneum. Adhesions bind together the serous surfaces with consequent disturbed function and movement as seen is case of traumatic pericarditis.

4. Organisation of fibrin in the alveoli of the lungs convert permanently the lung parenchyma into a fibrous tissue known as carnification.

3. Purulent Inflammation

In this, there is a formation of pus or purulent exudate i.e. a liquid of creamy colour. It contains fluid with large number of leucocytes, chiefly neutrophiles which are mostly dead due to action of toxins or harmful agents. Streptococci invite such neutrophilic reaction in the tissues. Turpentine or croton oil can also produce abscesses in the tissues. The pus contains peptones, polypeptides, aminoacids and enzymes of proteolytic character. In the purulent inflammation, liquefaction takes place in the focus of inflamed tissue as a result of which a cavity is formed and is filled with pus. Such localized or circumscribed collection of pus in a cavity is called an abscess and pus formed in this manner makes its way from the abscess to the exterior or into the internal cavities. Accumulation of pus in a closed cavity (serous cavity) or sinus is called empyema (i.e. for example, pus in pleural cavity or in the gall bladder). Purulent infiltration spreading diffusely through tissue spaces of loose inter cellular tissue or subcutaneous, muscular or interstitial tissue is called phlegmon (cellulitis). Pyaemia refers to formation of multiple abscesses due to entry of pyogenic bacteria into the circulation. Tissue destruction at surface in the skin and mucous membrane is called an ulcer which is marked by a breach in the surface of epidermal layer or mucous membrane. Pustule means the presence of pus on the skin and mucous membrane. Cellular exudate formed by parasites is dominated by eosinophiles. Nature of the inflammatory reaction is governed by the virulence of the infective agent and characteristics of the affected tissue. Pus may be thin and watery or inspissated and semisolid.

Pseudomonas aeruginosa causes formation of greenish pus. Pus from the black hooves of horses is of blackish discolouration due to sulphides. When considerable amount of pus is produced, the term suppurative inflammation is used.

Gross appearance

Viscous, cream coloured fluid fluid of watery consistency of thick yellow colour with foul small is formed.

Pus in the pulmonary alveoli is called grey hepatization due to greyish discolouration of the lung.

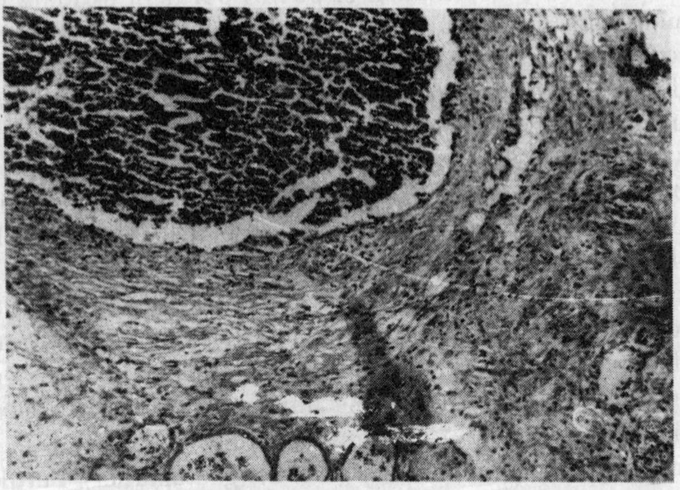

Fig. 23 Capsule of an abscess showing granulating fibrous tissue destroying and replacing the adjacent parenchyma in staphyloccal mastitis (H & E x 400)

Microscopic appearance

Pus consists of neutrophiles usually in large numbers and its presence in tissue confirms purulent inflammation (Fig.23). Neutrophiles may be dead with small, dark and irregularly shaped nuclei. Lymphocytes, plasma cells or macrophags may be present in varying numbers. Dead neutrophiles may have acidophilic cytoplasm and presence of hyperaemic or congestive changes in the inflamed zone is noticed.

Causes

1. Bacteria like *Corynebracterium pyogenes, C. renale, C, equi, staphylococci and streptococci.*
2. *Mycobacterium tuberculosis* is a pus former in the beginning and granulomas in actinomycosis, actinobacillosis, staphylococcosis, or botryomycosis are purulent in character.

Effects

1. Pus is due to liquefactive necrosis.
2. Toxaemia can occur in the affected individuals.
3. Infective organisms in the pus may give rise to metastases by entering the circulation and purulent focal growth in the tissues of organism seeded with these bacteria.

4. Haemorrhagic inflammation

In this, the exudate formed is rich in red cells and has a pink or pinkish red tint. Haemorrhagic exudate is formed in the inflammatory process caused by *Mycobacterium tuberculosis* as seen in pleurisy, peritonitis and pericarditis and *Bacillus anthracis* infection. Such exudate accumulates around the blood vessels in the inflamed zone. Red cells escape singly or in groups through large openings (called rhexis) in blood vessels whereas white cells (neutrophiles) pass through the cement lines of the otherwise intact capillary walls. There is an actually passive expulsion of the red cells i. e. a passage through the vessel wall by diapedesis. Haemorrhagic inflammation arises from avitaminoses, scurvy, inanition etc. and exuded red cells accumulate on the body surface or into neighbouring tissues. The exudate contains a large number of red cells along with fibrin, serum and leucocytes etc. and resemble clotted or unclotted blood.

Occurrence

1. It occurs in tissues in diseases like black leg, anthrax, pasteurellosis and purpura-haemorrhagica.
2. Found on the mucous surfaces.
3. Haemorrhagic gastritis and entritis are examples of haemorrhagic inflammation.

Causes

1. Highly virulent organisms.
2. Chemical poisons like phenol, arsenic and phosphorus. Mucous membranes come into contact with poisons and produce haemorrhagic exudate. Lining mucosa in the urinary bladder, gall bladder or lower bowel suffer from haemorhagic inflammation while excreting certain poisons through themselves.
3. Haemorhagic exudate is found in case of infectious laryngotracheitis.
4. Production of haemorrhagic inflammation by leptospirae and virus of Rubarth disease on account of injury to vascular endothelium.

Gross appearance

1. Presence of blood coloured material in either fluid or semifluid or in the clotted and gelatinous form on mucous surfaces and in tissue spaces.
2. Deep red inflamed surfaces
3. Presence of black coloured faeces. Posterior part of gastorointestinal tract may show haemorrhages.
4. Foamy or frothy blood escapes from nose in case of pulmonary haemorrhages.

Significance

1. It ends fatally.
2. Subsidence is rapid when the cause is removed.

5. Catarrhal or mucous inflammation.

It is characterized by excessive production of mucus by epithelial cells (mucous glands or one celled mucous glands i.e. goblet cells.) This kind of inflammation is limited to mucous surfaces. There is an excessive production of mucus which may become mucopurulent. Mucus may be thin, and watery in constistency. Even tubules of kidney may show mucus.

Causes

1. Mild irritants or chemicals acting over short period as formalin, chlorine, bromine etc.
2. Bacteria and viruses in cold. Mucous inflammation is marked in common cold in human beings.
3. Irritating foods in the digestive tract.
4. Violent or strong peristalsis causing transient diarrhoea.

Gross appearance

1. Clear, and slimy mucin containing fluid.
2. Drippings from the nostril due to inflammation of the nasal mucosae or increased flow of tears or saliva.
3. White shreds of mucus adhering to the formed faeces.

Microscpic appearance

1. Mucus visible as pale bluish or grayish strands of mucin clinging to the mucous membrane.
2. Increased number of goblet cells.
3. Desquamated or necrosed surface epithelium of the inflamed zone i.e. mucous membrane denuded of its epithelial lining.
4. Mucosa hyperaemic and some-what infiltrated with lymphocytes.

Effects

1. Mucus formation by the tissue is protective.
2. Mucous inflammation subsides with removal of the causes such as formalin and chlorine etc.
3. Hypersensitivity of cells in case of long action of the irritants like bacteria etc.
4. A chronic catarrhal inflammation follows.

OTHERS TYPES OF INFLAMMATION

Lymphocytic inflammation

It is characterized by an excessive accumulation of lymphocytes in tissues. Some hyperaemic changes may be present in the inflamed zone.

Occurrence

1. Central nervous system affected with different infections.
2. Islands of Glisson (portal spaces) in the liver.
3. Mucous membranes.
4. Kidneys as seen in subacute nephritis.

Causes

1. Bacterial or viral infections of the central nervous system such as rabies, equine encephalomyelitis and listeriosis etc.
2. Toxins causing inflammation like toxic hepatitis.

Gross appearance

It is not detected grossly.

Microscopic appearance

1. Lymphocytes are seen singly or in clumps in the tissues.
2. A wreath of lymphocytes around the blood vessels i.e. presence of perivascular cuffing as seen in the brain in some viral infections (e.g. equine encephalomyelitis and swinefever).
3. Infiltration of lymphocytes in the lamina propria of intestine.

Significance

1. Viral encephalitis with lymphocytic inflammation may terminate fatally.
2. Presence of lymphocytic inflammation concerned with mediation of immunologic responses.

Chronic inflammation (Productive inflammation)

It is also called granulomatous inflammation. The proliferative reaction takes place in the cells of the interstitum and inflammatory swelling partly arises from the proliferative changes. Chronic inflammation is a reaction to tissue products of the body itself such as clotted exudate, neutrophiles and red cells etc. The term chronic inflammation is also applied to those situations in which the reparative process is also visible and it can then be said to be including both the reparatory and regenerative changes.

The causes of chronic inflammation are as follows :

1. Protracted irritation to the tissue by noxious agents (e.g. bacteria, parasites and foreign particles etc.)
2. Products of decomposition and disturbed metabolism.
3. Low acidity (pH= 7.1 to-6.6) in the peripheral zone of the inflammatory area is conducive to the proliferation of cells like adventitious and reticular cells etc.)
4. Proliferation of endothelial cells of the blood and lymph capillaries, reticular cells, macrophages, plasma cells and lymphocytes appear in such zone.

Fig. 24 Actionomycosis (pharynx, cattle). A colony of *Actinomyces bovis* in a tissue surrounded by epithelioid cells and neutrophiles (H & E x 400)

Organization is a process by which dead tissue is removed by proliferating connective tissue cells and the defect is filled up. The newly formed granulation tissue is rich in blood vessels and grows into the defect i.e. it extends from periphery to the centre of inflammatory focus. The destroyed tissue is replaced and forms a demarcation line between the focus of inflammation and the healthy tissue. Finally, the connective tissue leaves a clear scar, in any injured tissue with tissue death. Inflammatory and reparative processes progress side by side i.e. there is no time gap between commencements of inflammation and repair in an injured tissue. Chronic inflammatory changes are seen in the pyogenic membrane or capsule of an abscess due to protracted bacterial activity. Neutrophiles are known to continuously walk through the pyogenic membrane into the lumen of the abscess. This membrane consists of granulation tissue (i.e. a tissue made up of newly formed fibroblasts and budding capillaries) arising from proliferating angioblasts of the capillaries). In short granulation tissue is depended on two processes, namely fibrosis and angiogenesis etc. filling in the defect or gap formed in the injured tissues.

Gross appearance of the granulation tissue

It is granular in skin and mucous tissue and such tissue growth fills up the defects. Its appearance is like pyogenic membrane. This later develops into a scar formed at the site of defect. Presence of rounded nodules (capillary buds) or granulations is seen in the newly formed growth.

Microscopic appearance of the granulation tissue

There are numerous netrophiles which have passed out from the deeply situated capillaries in pyogenic membrane adjacent to the lumens of the abscesses. Just next to this layer, monocytes and histiocytes i.e. the cells wandering in such growths from the blood and peripheral connective tissues, are seen.

Such cells can be seen accumulated around the blood vessels. Newly formed capillaries arising or resulting from the budding processes of endothelial cells or angioblasts from the pre existing capillaries of the inflamed zone can be found in the pyogenic membrane. The newly proliferated zone can be found in the pyogenic membrane. The proliferated cells i.e. fibroblasts from the pre existing interstitium or connective tissue of the inflamed zone can be found in the pyogenic membrane. The proliferated cells i.e. fibroblasts etc. are arranged around these capillaries and newly formed fibroblasts from the pre-existing connective tissue cells follow the new capillary buds and the fibres formed by these cells follow the new capillary buds and constitute a supporting network. The tissue towards the periphery of the pyogenic membrane become more fibrous. The mesenchyme behaves in the same way as in the case of many chronic inflammations.

Mesenchymal proliferation is also seen in the liver, kidneys and bladder etc. Small nodules are visible in the contracted firm tissues. Parenchymal cells in such organs show degenerative changes and there is a perivascular accumulation of lymphocytes, plasma cells and histiocytes. Such changes are chronic inflammatory changes.

Chronic inflammation actually arises from protracted inflammation due to weak stimulus. The local tissue cells of chronic inflammation result from mild and continuous irritation from the outset. Non-living substances like COA particles or toxins of bacterial or metabolic origin set up chronic inflammatory changes. Organisms such as *Mycobactrium tuberculosis, Actinomyces bovis, Actinobocillus ligniersi and Pseudomonas mallei* show proliferative changes to produce infective granulomas or infective granulomata. (Fig. 24) The term granuloma refers to a tumour like lesion of granulation tissue.

Microscpically, the changes in chronic inflammation are characterized by proliferation of the connective tissues and endothelial cells in the inflamed focus. The connective tissue proliferates to give rise to fibroblasts which form fibrous tissue (Fig.25). Proliferation of histiocytes, macrophages and cells of endothelial organs also occurs. Sutures, splints or oil give rise to foreign body granulomas which consist of macrophages, some neutrophiles and multi-nucleated giant cells (foreign body giant cells). In these giant cells, nuclei are randomly distributed. Tuberculous granulation tissue in organs is marked by proliferation of epithelioid cells, formation of Langhan's type of giant cells, new blood vessels and presence of proliferated fibroblasts etc. (Fig. 26-27.)

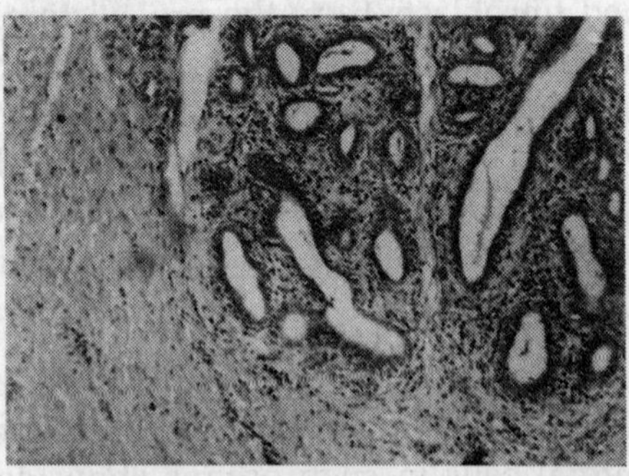

Fig. 25. Widening of the interductal and interalveolar interstitium with marked fibrosis and infiltration of round cells in chronic mastitis in buffalo (H & E x 100.)

In case of tuberculosis and fungal infections, granulomas mainly consist of macrophages or epithelioid cells (cells with appearance like epithelial cells). Fusion of these macrophages gives rise to Langhan's type of giant cells. Granulomas are surrounded by immature connective tissue which is infiltrated by lymphocytes. These lesions, later, become scars.

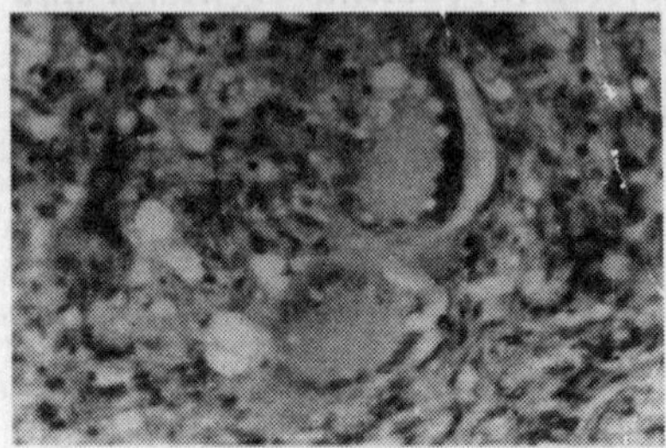

Fig. 26 Epithelioid granulation tissue in tuberculous mammary gland of a buffalo.Note the epithelioid cells, Langhan's giant cells, lyphocytes and fibroblasts (H & E x 400).

Fig. 27 Epithelioid granulation tissue (lung,cattle) showing Langhan's giant cells and epithelioid cells(H& E x 400)

Granulation tissue is considered as a distinct aspect from the new growth of granulomatous inflammation. In this, the lesions show red granules (formed by blunt ends of the new capillary loops). It is seen in wounds or sinuses. Grossly, the surface of the lesion bugles, shows granules under the scab (due to capillary loops) and the tissues in deeper part are white or tough. Its exposed surface may be ulcerated and bloody. Microscopically, there is a presence of fibrillar cells with plump, ovoid and hyperchromatic nuclei. The deeper parts of the lesions show bundles of fasciculi. Many new capillaries have plump endothelial nuclei and the capillaries are arranged perpendicularly to the surface. Neutophiles, red cells and tissue debris are seen in the zone nearest to the surface.

Summing up, chronic inflammation shows two kinds of important lesions, namely.

1. Granulomas (reticulo-endothelial granulation tissue)
2. Granulation tissue (fibrous or pyogenic granulation tissue).

Reticuloendothelial proliferation, thus, is a feature in the granulomas (infectious).. e.g. a tubercle in *M.tuberculosis* infection whereas fibrous granulation tissues is recognized due to the presence of newly formed fibrous tissue and numerous young capillaries filling in the defects in the injured tissues.

The main distinguishing features between acute-and chronic inflammations are given in Table-7

Table-7 Some features of acute-and chronic inflammations

Acute inflammation	Chronic inflammation
1. Sudden onset often within a few minutes or hours because of a strong stimulus. Limited or short lived inflammatory changes. Irritants usually removed in short-lived inflammation.	1. A longer duration (i.e. more time taken in terms of weeks or months) and infiltration of plasma cells, lymphocytes and monocytes in the inflamed area. Persistence of the irritant in long lived chronic inflammation .
2. Presence of clinical signs of inflammation e.g. signs of redness, swelling, heat, pain and functiolaesa (impaired function) as a result of tissue injury .	2. No increased temperature and red colour or pain in inflamed zone .

3. Morphological signs of degenerative, exudative and inflammatory changes. Formation of exudate as a characteristic change in the inflamed area.	3. Prolongation or progression of acute inflammation into chronic inflammation (i.e. acute-to-chronic transition). Presence of chronic inflammation from the very outset of inflammatory changes. Morphological signs of degenerative, exudative and inflammatory changes are noticed in chronic inflammation (still active).
4. Red, serous, purulent and fibrinous exudates as bases of classifying acute inflammation	4. ————
5. Presence of vascular changes e.g. vasodilatation, increased blood flow followed by slowing of blood stream, stasis, endothelial injury, increased vascular permeability and escape of proteins in the extravascular space. Complete restoration of acute inflammation to normal (i.e. resolution) and return of normal vascular permeability, cessation of leucocytes (largly neutrophiles) infiltration and removal of oedema fluid, foreign agents, pathogenic microbes, protein, leucocytes and necrotic debris etc. from the inflamed region.	Revelation of lymphocytes monocytes, and fibroblasts in abundance by light microscopy (LM). Attempts at healing marked by connective tissue replacement (fibrosis), proliferation of small blood vessels (i.e. capillaries and venules etc. called angiogenesis). Presence of less complete resolution and less frequently presence of regenerated tissue in organs like liver. Indication of filling by fibrous tissue replacement of the dead or destroyed tissue. Growth of fibrous tissues in the zone of exudate or fibrinous clot i.e. the processes of organization. Secretion of cytokines like INE -γ by sensitized T lymphocytes, activation of macrophages by these proteins and phagocytosis and destruction of pathogenic microbes by activated macrophages. Infiltration of leucocytes caused by TNF-α at the inflamed foci.

Subacute inflammation

Some pathologists still use the term subacute inflamation which is an inflammatory reaction positioned between acute and chronic inflammations in tissues or organs .

Its chief features are as follows:

1. Less explosive and less intense inflammatory reaction caused by a weak stimulation with a long period of its activity (say, a few weeks).
2. An intermediate reaction between extremes of acute- and chronic inflammations.
3. A sequel of an acute inflammation or subacute type from its very outset in tissues or organs.
4. Less distinct presence of swelling, redness, pain, and temperature in the inflamed tissues.
5. Infiltration of leucocytes e.g. a few neutrophiles, some eosinophiles and more plasma cells and leucocytes in the inflamed foci.
6. Presence of immature connective tissue in the lesions as a very characteristic change. Acute inflammation differs from subacute or chronic inflammation because of no proliferative changes in cells e.g. fibroblasts in the inflamed area. Immature proliferative tissue and mature fibrous tissues in the inflamed area are features of subacute and chronic inflammations respectively.

Inflammatory mediators of inflamed tissues in organs

Apart of from several animate and inanimate inflammatory agents, there are several chemical substances or products which participate in the inflammatory reaction. Many of the inflammatory mediators in the inflamed tissues are derived from the plasma exuding from the blood vessels with important enzyme systems, namely (1) clotting system, (2) kinin system (3) fibrinolytic system and (4) complement system. Some details of inflammatory mediators are given in Table-8.

Table-8 Some, main roles of inflammatory mediators e.g. vasoactive amines' arachidonic acid (AA) metabolites, cytokines and lysosomal enzymes etc.

Molecules / Inflammatory mediators/ Cytokines	Sources	Actions/ Roles
1. Histamine and serotonin (vasoactive amines released from granules following proper stimulation)	Mast cells, basophiles and platelets storing these amines in the secretary granules	Dilation of arterioles and capillaries, increased vascular permeability, vascular leakage and contraction of smooth muscle by histamine
2. Bradykinin (generated from digestion of high molecular weight kinonogen (HMW-kinonogen) by the enzyme kallikrein)	Plasma substrate	Pain and vascular leakage, increased vascular permeability, vasodialation, endothelial contraction, widening of endothelium gaps, activation of enzyme phosphorlipase A_2 and liberation of archidonic (AA) metabolites
3. Complement components	Complement components (i.e. plasma derived factors)	———
a. C3 a (an anaphylatoxin)	Plasma proteins via liver	Release of AA metabolites from the phospholipids in the lipid bilayer of the cells membrane, increased permeability, vascular leakage and derangulation of mast cells
b. C3b	A spilt product of C3 by C3 convertase	Acting as an opsonic fragment
c.C5a	Macrophages	Increased vascular permeability, vascular leakage, chemo tactic factor for leucocytes, adhesion and activation of leucocytes and deregulation of mast cells.

d. MAC (membrane attack complex- C5-9	DO	Lysis of foreign cells and pathogenic bacteria etc.
4. AA metabolites a. Prostaglandins	Mast cells and phospholipids in the cell membranes etc.	Pain, vasodilatation and fever by PGE $_2$ and PG1$_2$ (prostacyclin)
b. Thromboxane A_2 (TXA_2)	Phospholipids in cell membranes i.e. activated platelets	Pulmonary vaso constriction, chemotaxis platelets aggregation and adherence of neutrophiles to endothelium
c. Leukotriene C4, D4 and E4 (also called SRS- A- slow reacting substances of anaphylaxis)	Mast cells and leucocytes	Vasoconstriction and bronchoconstriction
d. Leukotriene B4 (LTB4)	Leucocytes	Degranulation, digestion and production of reactive oxygen species or toxic intermediates, phagocytosis, chemotaxis and activation and adhesion of leucocytes
e. Lipoxin A and B	DO	Chemotaxis (directed movement of leucocytes towards chemoattractants and stimulation of granulocytes by lipoxin A.
5. Oxygen metabolites (O_2,-), H_2O_2 and OH·)	Leucocytes	Damage to endothelium, tissues , lipids proteins and nucleic acid in cells. Destruction of pathogenic bacteria at the inflamed site.
6. PAF (platelet activating factor or AGEPC (actyl- glyceryl- ether-phosphoryl choline)	Leucocytes and mast cells	Bronchoconstriction, increased vascular permeability and vasodilatation, chemotaxis, stimulation of leucocytes and induction of local inflammatory cells
7. Cytokines IL-1 (interleukin-1), IL- 8 and tissue necrosis factor-α (TNFα) Proteins influencing the activities of other cells	Activated macrophages	Chemotaxis, tissue damage, endothelial cells activation and acute phase reaction. Mediation of gram-negative shock in the body of patients by TNFα

8. Chemokines (chemical mediators) MCP-1 (monocyte chemoattractant protein, eotaxin and macrophage inflammatory protein-k (MIP-K)	Leucocytes	Chemotaxis and activation of leucocytes e.g. eosinophiles, basophiles and lymphocytes. Recruitment of eosinophiles by eotaxin.
9. Nitric oxide (No)	Endothelium, neurons and macrophages	Vasodilatation and cytotoxicity to cells or tissues and bacteria.
10. Integrins (leucocyte receptors for adhesion molecules e.g. ICAM-1 and VCAM-1 VLA (very late antigen proteins) like VLA-1, 2,4,5 and 6.	Activated monocytes	Chemotaxis and interaction of the adhesion receptor of molecule ß1 integrins (VLA proteins). & ß2 integrins (including LFA-1, Mac – 1 and CD 11_a/ CD-18) with VCAM-1 and ICAM-1 respectively. Adhesion and transmigration of leucocytes caused.
11. Plasmin	A protease as an inactive precursor called plasminogen in the globulin fraction of the blood	Degradation of fibrin to soluble fibrin split products to be drained away by lymphatics, initiation of the the conversion of HMW-kinonogen to bradykinin
12. Neuropeptides e.g. substance-P and neurokinin A	Present in the central and peripheral nervous tissue	Initiation of inflammatory process, increased vascular permeability, capillary leakage in lungs and intestines, pain signals and regulation of blood pressure.
13.Lysosomal enzymes a. Lysozymes collagensase,(a protease), gelatinase histaminase, alkaline phosphatase and acid hydroleses	Granules of neutophiles Azurophil granules of neutrophiles discharge their contents into phagolysosome (phagocytic vacuoles) contaning the ingested pathogenic bacteria	Increased vascular permeability, chemotaxis and tissue damage produced by lysosomal constitutents of leucocytes

b. Acid hydrolases, collagenase, elastase (protease) phospholipse and phosphatase	Monocytes and macrophages Half life of blood monocytes - About a week Life span of macrophages- Several months Macrophages are converted into activated forms i.e. epithelioid cells and giant cells. Kupffer cells (liver), septal or alveolar cells (lungs) and osteoblasts (bone arising from differentiation of monocytes	
14. Proteases (enzymes hydrolyzing peptide bonds of poly peptides) Plasmin, Kallikerin and globurin permeability factors are example of prtoteases	Plasma derived factors (proteases as in the system of complements, kinin and clotting etc.	
a. Acid proteases (acting at an acid pH).		Digestion of bacteria and cellular debris
b. Neutral proteases		Degradation of extra cellular proteins and cleaving of C3 and C5 to release anahylatoxins

Chemical mediators of inflammation

The chief chemical mediators are :

1. Histamine (Bimidazolyl-ethylamine)

It initiates the early vascular responses and sustains them for 30 to 60 minutes. Later, other mediators participate to maintain the vascular reactions or produce the delayed or prolonged vascular reaction. This substance is present in the granules of basophiles and mast cells which release histamine owing to injury to the cell or tissue in the organ. Local vasodilation (i.e. dilation of the minute vessels like capillaries) producing the central dull red area in the tissue after injury, arises due to action of histamine or histamine like substance. This is the first stage of the triple reaction as described by Lewis (1927). Histidine decarboxylase, an enzyme, takes part in the formation of histamine is known to increase in the state of hypersensitivity or burn cases.

Serotonin (5-hydroxytryptamine)

It causes vascular dilatation and increased permeability in the injured tissue and is found in the platelets, argentaffin cells of the gastro-intestinal tract and central nervous system. It is also present in the stings of many animals and plants. This substance is a pain producer, causes capillary permeability in the rat skin and is responsible for a flare on intradermal injection in man. The reaction of flare is the second stage of triple reaction or response.

Kinins (Bradykinin and kallidin)

The vascular reactions after the initial hisamine response are sustained by kinins. Bradykinin is a very strong vasodilator and causes increased permeability of the capillaries of guinea-pig and rat skin.

Prostaglandins

These small fatty acid derivatives affect many inflammatory reactions and are important for chemotaxis. P.G.E1 and P.G.E 2 mediate increased vascular permeability, potentiate the effect of histamine and bradykinin causing pain and also act as pyrogens. These are produced from fatty acids and released from phospholipids during cell injury.

Other permeability factors and slow reacting substance of anaphylaxis i.e. leukotriene C4, D4 and E4

Globule permeability factor, a lymph node permeability factor, hyaluronidase, lecithinase and necleosides are also permeability factors but their role is not clear. Slow reactive substance of anaphylaxis (S,R,S,A,), an acid sulphur containing lipids, are released from sensitized cells and causes increased vascular permeability and contraction of smooth muscles. Lactic acid is produced by emigrated leucocytes. It increases permeability of the small blood vessels. Substances like epinephrine and norepinephrine which cause vasoconstriction and diminish vascular permeability are anti-inflammatory substances.

Alterations in the focus of inflammation in vascularised tissue

Alterative tissue changes (cellular swelling and fatty changes etc.) in the inflamed focus is due to factors like disturbances in the metabolism, infections and intoxications. There is an increased or diminished vital function shown by the cells. A weak irritation causes division of the cells. Strong irritation diminishes the metabolism of the tissue and can lead to necrosis. Collagenous and elastic fibres may swell and disslove in the intercellular substance. Traumas, burns, strong acids and alkalies cause necrotic changes. Oxidative process is diminshed in the centre of inflammantory focus, but in the other parts, these may increase. Anaerobic glycolysis is noticed in the inflamed tissue. In the inflamed tissue, carbohydrates are not fully oxidized leading to accumulation of lactic acid in the tissues. Fatty acids, and ketone bodies are formed in the focus of inflammation due to disturbances in fat and protein metabolism. Acidosis arises from accumulation of acid metabolites and concentration of hydrogen ions increases in the inflamed tissue. The pH in such tissue is 7.1 - 6.6 in chronic inflammation and 6.5 - 5.39 in acute purulent inflammation respectively. The concentration of potassium ions increases in the inflamed tissue. Osmotic pressure in the extravascular areas increased due to excess of ions and products of tissue disintegration but it gradually diminishes towards the periphery of the focus of inflammation.

Thus, the four main changes in the inflamed tissues are as under:

1. Dystrophic changes like cellular swelling, fatty changes and necrosis.
2. Accummulation of ions like hydrogen ions.
3. Increase in colloid osmotic pressure.
4. Exudative, emigrative and proliferative changes.

To sum up, some details of chief different mediators which

influence inflammatory processes in the injured tissue are :

1. Vasoactive amines or peptides e.g. histamine and serotonin etc.
2. Arachidoic acid (AA) metabolites.
3. PAF (platelet activating factors).
4. Oxidising agents or reactive oxygen species.
5. Plasma proteases.
6. Chemokines.
7. Neuropeptides

Histamine

Degradation of preformed histamine in mast cells granules releases amines like histamine.

The chief causes of histamine release are :

1. Physical factors e.g. heat, cold and trauma etc.
2. Immune reactions characterized by binding of antibodies to mast cells.
3. C3a and C5a (fragments of the complement).

These fragments cause anaphylaxis in the body.

4. Cytokines like IL-1, IL-8 and neuropeptides e.g. substance P.

The effects of histamine released are :

(i) Arterial dilation increased permeability of the venules and contraction of large arteries.

(ii) Widening of interendothelial junctions or gaps causing increased vascular permeability.

Histamine acts on microcirculation comprising of pre-capillaries, capillary bed, post-capillaries and post-capillary venules.

Serotonin

It occurs in mast cells and plateles as preformed mediators and acts like histamine and platelate activity factors.

PAF causes aggregation of platelets and release of serotonin. Thrombin and collagen cause aggregation of platelets following their contacts with platelets and a release of histamine and serotonin is noticed from platelets.

Complement system

There are 20 components of compliment in plasma. Membrane attack complex (MAC i.e. C5-9) causes lysis of bacteria or cells. The enzyme C5 convertase interacts with C5 to release C5a. The assembly of MAC is initiated by C5 convertase. C3a, C5a, C4a, act like anaphylatoxins which cause increased vascular permeability, vasodilation and release of histamine from mast cells. The lipoxygenase pathway of AA metabolism is activated by C5a. C5a also activates leucocytes and causes their adhesion to endothelial cells. It also increases avidity i.e. (binding affinity in sense) of surface antigens to their complementary endothelial adhesion molecules. Lysosomal enzymes activate complement components like C3 and C5. C3b and C3bi act like opsonins which coat the bacterial cell walls and facilitate the phagocytosis of the opsonin coated particles by neutophiles and macrophases. C1 inhibitor and CD_{59} show regulatory effects on the activity of complement of components. And as such, one notices destruction of pathogenic bacteria and prevention of tissue injury by activated complement system.

Kinins

Some espcific enzymes i.e. proteases called kallikerins elaborate vaso active peptides (bradykinin) from plasma proteins known as kininogenes e.g. HMKW kininogenes (high moleculer weight kininogenes). Bradykinin increases vascular permeability, widen interendothelial gaps. It also stimulates phospholipase A2 which liberates archidonic acid from the phospholipids in the lipid bilayer of the cell membrane. High concentration of intracellular calcium activates the enzyme phospholipase A2.

Arachidonic acid (AA) metabolites

Activated leucocytes, platelets and mast cells librate vasoactive AA metabolites. Increased level of cytoplasmic calcium activates phospholipase A2 which causes hydrolysis of arachidonic from the cell membrane. Phopholipase C and dicylglycerol lipase play important roles in the release of AA metabolites from cell membranes and casuses release of the AA platelets. Perturbation of the lipid bilayer of cell membrane by complement activation also causes release of such metabolites from cell membranes. Bradykinin and TNF-α cause increased level of intracellular cytoplasmic calcium. The metabolic products of arachidonic acid via cyclo-oxygenase pathway are prostaglandins and thromboxane A2 (TXA2) whereas leukotrienes and lipoxins are the metabolic products via lipoxygenese pathway. PGD2, PGDE2 and $PG1_2$ (prostacoyclian) are examples of prostaglandins. Prostocyclin is elaborated by endothelium. Activated platelets generate thromboxane A2. Prostacyclin causes vasodilatation, inhibition of plateletes aggregation, increased vascular permeability and chemotaxis. Vasoconstriction, promotion of platelets aggregation and adherence of neutophiles to endothelial cells are produced by thromboxane A2. Prostaglandins like PGD2 and PGE2 show vasodilatory property. Prostaglandins and leukotrienes act synergistically to influence vascular permeability and inflammatory reaction

Cytokines

These are proteins elaborated by activated leucocytes, macrophases, endothelium, epithelium and connective tissue cells etc. The cytokines play important roll in acute achronic reaction. Interlukins are cytokines acting on leucocytes. Chemokines are also cytokines which stimulate leucocyte movement (chemokinesis) and cause directed movement of leucocytes i.e. chemotaxis. Certain growth factors (GF) also act like cytokines. IL-1 to IL10, TNF-α, interferon α, MCAF (monocytes chemotactic and actvating factors) and M-CST (microphages colony stimulating factor) are examples of certain important cytokines. Autocrine signaling of cytokines is marked by its effects on the target sites on the same secretory cells. Paracrine signaling is marked by an action of the cytokine released on the adjesent target cells in the vicinity of the secretary cells. But an endocrine action of a cytokine is one in which the effect of the cytokine is noticed on some distant cells through blood circulation.

Inflammation in non-vascular tissue

Changes in cornea due to injuries are not so severe as seen in the injured vascularised tissue. Reparative activity is shown by proliferation of local fixed connective tissue cells in the absence of vascular tissue changes. Some vascular changes like exudation and emigration of leucocytes are found at the periphery of the cornea in extensive injury to it by irritants like zinc chloride.

Outcomes of inflammation

1. Subacute or chronic type may arise as a sequel from acute inflammation.
2. Vasoexudative changes in acute inflammation are predominant phenomena.
3. Protracted course in chronic inflammation is seen with mild symptoms.
4. Return to normalcy with restoration of anatomic and functional properties.
5. Formation of scar.
6. Displacement of organs or functional disturbances as seen due to scars or contractures in the central nervous system.
7. Destruction of tissue and even death of the affected organisms.

Leucocytes in inflammation

Vascular exudative, emigrative and defensive changes take place in the irritated tissues on account of noxious agents. According to some authors, vascular and exudative changes are secondary changes and changes like cellular swelling, fatty changes and necrosis etc. are primary changes. Role of inflammatory cells in the injured tissue is extremely important and directed towards defence of the body from further injurious effects. It is worthwhile to note that the parenchymal cells or the component cells of an organ are the first targets of injurious agents e.g. hepatocytes and renal epithelium to phosphorus.

Neutrophiles (Microphages)

The main roles are given below :

1. Present in large numbers in the purulent exudate (pus).
2. Emigration from blood vessels to the site of inflammation due to their amoeboid movement with the object of engulfing the harmful cocci and smaller bacilli.
3. Secretion of lytic substances (lysosomal enzymes) to dissolve the dead bacteria and dead cells in the inflamed zone. Dead neutrophiles are replaced by fresh cells from the myeloid tissue in the bone marrow.
4. Increase in the percentage of neutrophiles in the acute pyogenic infection to destroy the pathogens. There is also a decrease in the white cell count in some viral diseases. (e.g. rinderpest).
5. Young or juvenile or also stab forms of neutrophiles present upto 6% of the total white cells in pyogenic infections. The term shift to the left refers to an increase in the count of juvenile forms.

Monocytes (Macrophages)

These cells are seen in large numbers in the inflamed zone after lapse of a few days following the onset of inflammation. They possess motility and are capable to phagocytise large particles of animate and inanimate origin. Monocytes fuse to form giant cells of two types, which are as follows:

1. Langhan's giant cells (about 60 microns in diameter) with spherical nuclei arranged in a more or less complete wreath form just inside their hazy or indistinct borders.
2. Foreign body giant cells (recognized by presence of larger nuclei in the form of a jumbled mass at the centre of the cells.) These cells secrete enzymes and neutralizing substances. Migration of monocytes from the circulation and mitotic proliferation of the related cells of the mesenchymal origin add to an increase of such cells in the inflamed zone.

Eosinophiles

These cells appear in the inflamed tissues in the condition of antigen antibody reaction (e.g. in tuberculin test). Some other features of eosinophiles are the following :

1. Present in large number in the local tissues in parasitic infections and allergic reactions.
2. Eosinophilia refers to an increase in the number of eosinophiles.
3. These cells show some amoeboid movement and are presumed to engulf and destroy particulate matters such as dead cells, bacteria and parasites. Eosinophilic extracts are antagonistic to histamines, serotonin and are thought to be physiologic opposites of the mast cells.

Eosinophiles, with or without other leucocytes, accumulate at the site of allergic reactions and enter into the viscinity of animal parasites.

Basophiles

Basophiles and mast cells have similar function and contain histamine and heparin. There is little knowledge about the activities and properties of the basophiles.

Lymphocytes

These cells occur in inflamed tissue in both acute and chronic inflammations. They are known to possess some amoeboid movement. When they gather around blood vessels in large numbers, the phenomenon is called perivascular lymphocytic infiltration in the form of a cuff. Neutralizing and lytic enzymes are produced by these cells to possibly destroy toxic material of protein metabolism. Removal of thymus or bursa from the embryo before the maturity of the lymphocytic system produces immunological defects. The lymphocytes, i.e. T and B lymphocytes are morphologically similar but functionally distinct. Lymphocytes can be divided into small lymphocytes and large lymphocytes. Small lymphocytes can also differentiate into the larger types of lymphocytes.

Plasma cells

These are modified lymphocytes capable to produce gamma globulins. The cells are spherical or elliptical with nuclei having a clock face peripheral arrangement of chromatin granules. The nuclei in the plasma cells are always eccentrically placed in the cells and the cytoplasm is little more basophilic in staining character and has a life span of about 12 hours. Small plasma cells can be seen in almost all inflammatory reactions, particularly in inflammation of the intestines and male reproductive organs.

Mast cells

These cells arise from undifferentiated mesenchymal cells by the process of mitosis. They are found in loose connective tissues in the mammary glands. These cells contain many spherical basophilic and metachromatic granules which contain heparin and histamine, sertonin (in the rat and mouse) and also different kinds of enzymes. They can sometimes grow to form tumours.

Summarising, inflammation is a fundamental protective reaction with a purpose in an organism. Repair a concurrent phenomenal with inflammation, and fibrosis is marked by disfiguring scars, fibrous bands or adhesions limiting the mobility of certain organs (e.g. heart in traumatic pericarditis in bovines) and even obstructions e.g. some kinds of intestinal obstruction The inflammatory response in vascularised connective tissue is marked by activities of certain cells as given below :

1. Circulating cells e.g. neutrophiles, monocytes, lymphocytes, eosninophiles and platelets.
2. Connective tissue cells e.g. fibroblasts, mast cells, lymphocytes and macrophages.

It is better to describe the extracellular constituents of connective tissues in the injured areas which are the following.

1. Structural fibrous proteins (collagen and elastin).
2. Adhesive glycoproteins (fibronectin, laminin and nonfibrillar collagen etc).

3. Basement membrane (BM) – a specialized part of the extracellular matrix. It consists of adhesive glycoproteins and proteoglycans.

Inflammatory mediators e.g. chemical factors derived from plasma (i.e. the complement system), mediate the vascular and cellular responses in both acute and chronic inflammations. Necrotic cells generate inflammatory mediators. Inflammatory reactions disappear from the scene of injury when noxious stimuli are removed.

Some important features of acute and chronic inflammations are as follows:

1. **Acute inflammation**
   a. Exudation of fluid or plasma proteins as noticed in the inflammatory oedema.
   b. Emigration of leucocytes, that is, neutrophiles and monocytes.
   c. Duration of acute inflammation in the injured tissues. It extends over a very short time (say , a few minutes to some days).

In the inflamed tissues, plasma exuding from the blood vessels bring with it important plasma enzyme systems, namely: (1) Coagulaton system, (2) Kinin system, (3) Fibrinolytic system and (4) Compliment system.

2. **Chronic inflammation**

a. Inflammatory cells e.g., lymphocytes and macrophages

b. Proliferation of endothelial cells to form new blood vessels (called angiogenesis), fibrosis (proliferation of fibroblasts) and degenerative changes e.g. necrosis of tissues in the inflamed zone.

c. Duration of chronic inflammation.

It lasts over a longer period in the injured tissues i.e. for weeks and months.

Cellular events in the inflamed tissues

The chief cellular events (i.e. an involvement of leucocytes) are:

1. Leucocytes extravacation.
2. Phagocytosis.

Leucocytes extravasation

It means the sequence of events in the movement of leucocytes from the lumens of blood vessels into the interstitial tissues in the inflamed zone of organs. Delivery of leucocytes to the sites of the injuries leads to ingestion of pathogenic, microbes, degraded necrotic cells and foreign substances e.g. antigens. Leucocytes in the inflamed zone release enzymes, chemical mediators and reactive toxic intermediates (free radicals) which even lengthen the inflammatory reactions in the injured tissues.

The chief steps of extravasations of leucocytes are:

1. Margination, rolling and adhesion of leucocytes
2. Transmigration of leucocytes across the endothelium (i.e. diapedesis).
3. Migration of leucocytes in the intestinal tissues.

The emigrated leucocytes in the injured tissues move towards chemotactic stimuli. Red cells which are normally seen in the central axial column of the formed elements of the blood displace the leucocytes towards the endothelial lining of blood vessels. Thus, laminar blood flow maintains the leucocytes against the vascular wall. A slow blood flow is seen in the blood vessels of inflamed tissues due to increased permeability with consequent outflow of serum into the extravascular spaces or intercellular spaces. Then, white cells adopt a peripheral position along the vascular endothelium. Such accumulation of leukocytes along the vascular endothelial cells is called margination of leukocytes. The leukocytes tumble along the endothelium and adhere to it transiently. This slow tumbling of the leukocytes along the endothelium of the blood vessels is called rolling. (Table 6) The leukocytes adhere firmly to endothelium of the blood vessels and behave like pebbles along the endothelium. Such phenomenon of adhesion arises from a complementary binding of the endothelial adhesion molecules to receptors on the leucocytes (i.e. a ligand-receptor interaction). There is a virtual lining of endothelium with leukocytes called pavementing by leukocytes. Then, the leukocytes penetrate through the junctions or gaps between the endothelial cells of the blood vessels by inserting their pseudopods and finally squeeze or escape through the interendothelial junctions. The leukocytes emigrate through the basement membrane (BM) and enter into the extravascular spaces . All the leukocytes like neutrophiles and monocytes etc. follow a similar path for extravasation into the extravascular tissues spaces.

Chemical mediators and some proteins *e.g. chemoattractants and cytokines.*

These inflammatory mediators modulate the surface expression or avidity of adhesion molecules produced by vascular endothelium. The endothelial cells activated by cytokines like IL-1 and TNF-α undergo hypertrophy, became more plump and generate adhesion molecules like EL selectin (ELAM -1), ICAM-1 and VCAM-1 etc. These endothelial cell membrane proteins mediate leucocytes adhesion to the endothelial surface. The important adhesion molecules include (1)E-selectin (2) Immunoglobulins (3) Integrins (leukocyte receptors) and mucin glycoprotein. (Table 6)

There are certain activities prior to binding of complementary adhesion molecules on the endothelial cells and receptors on leucocytes.

The chief steps of such activities are:

1. Redistridution of adhesion molecules.

2. This is a marked by redistribution of the adhesion molecule on the cell surface of vascular endothelium. P-selectin i.e. an adhesion molecule is normally *present* in the membrane of intracytoplasmic endothelial granules called Weibl-Palade bodies. Inflammatory mediators like histamine, thrombin and platelet activity factor (PAF) stimulate the endothelial cells. As

a result, P-selectin is rapidly distributed to the endothelial cells surface. Then, these adhesion molecules bind to receptors on the leucocytes. Process of rolling seems to arise from such receptor- ligand binding. IL-1 and TNF- α induce surface expression of endothelial adhesion molecule (e.g. E-selectin) which binds to receptor on the neutrophiles, monocytes and lymphocytes resulting in adhesion of endothelial molecule like E-selecting to vascular endothelial cells such adhesion of leucocytes to endothelial is noticed in about 1 to 2 hours of initial cell injury to tissues.

3. Integrins and their increased avidity. (binding affinity in sense)

Chemotactic agents (chemokinins) activate the leucocytes. When leucocytes like neutophiles, lymphocytes and monocytes etc. are activated by chemokines (e.g. IL-8 for neutophiles, MCP-1 – monocytes chemoattractant protein-1 for monocyte and lympohtoxins for lymphocytes) , β1 integrin (LFA-1) on the leucocytes is converted from the state of low affinity to high affinity to combine with ICAM- 1 (an intercellular adhesion molecule). This leads to firm adhesion of leucocytes to vascular endothelium (e.g. ICAM-1 /LFA-1 binding). This binding process of complementary molecule on the endothelial and leucocytes surfaces is followed by subsequent emigration of leucocytes across the endothelium.

In short, rolling is induced by binding of E or P selectin on endothelium on receptors like PSGL-1 and ESL-1 on the leucocytes. Chemokines activate the leucocytes and enhance avidity of α2 integrins for ICAM-1 expressed on the endothelial cells. PECAM-1 is an adhesion molecule present in the cellular junction of the vascular endothelial cells and assists in transmigration of leucocytes across the vascular endotheliam into the extravascular space. Neutophiles are predominantly present in the infiltrate during the first six to twelve hours but these are replaced by monocytes in 24 to 48 hours. Neutrophiles die and disintegrate within 24 to 48 hours. But monocytes in the infiltrate of the inflamed tissues live there for a longer period. In infections of staphylococci and pseudomonas, one sees predominant presence of neutophiles in the exudates. Viral infections are marked by infiltration or accumulation of lymphocytes around blood vessels (e.g. perviscular cuffing) where esionphiles are dominant cells in allergic, parasitic and hypersensitive reactions.

Chemotaxis

This is a very important part of an inflammatory phenomenon in the injured tissues. The leucocytes adherent to endothelium of vascular channels because of interaction between endothelial adhesion molecules and receptors on leucocytes transmigrate through the vascular endothelium into the extravascular areas. This process of extravasation is followed by directed movement of leucocytes by chemoattractants towards the site of injury along a chemical gradient. Neutophils, monocytes and lymphocytes respond to chemotactic stimuali exerted by chemoattractants.

The chief chemoattractants are:

1. Exogenous materials.g. bacterial products or chemicals.
2. Endogenous materials or mediators.

The main endogenous mediators are :

1. Complement components e.g. C_5a.
2. Products of lypoxygenase pathway e.g. leukotriene B_4 (LTB_4).
3. Cytokines e.g. IL-8.

The chemoattractants combine with specific receptors on the cell membrane of leucocytes. The interaction between chemoattractants and receptors on the leucocytes causes activation of phospholipase C. This reaction is followed by release of calcium from intracellular stores e.g. endoplasmic reticulum. Even influx of extracellular calcium occurs in the leucocytes. The high concentration of cytosolic calcium causes the assembly of contractile elements or filaments e.g. actin and myosin etc. which culminate in to motality or movement of leucocytes.

Phagocytosis

It denotes the act of engulfment of objects like microbes and foreign particles, forgein red cells and carbon particles etc. Leucocytes like neutrophiles are called microphages whereas monocytes are examples of macrophages. Organisms like strepto and stlaphylococci and pneumococci are engulfed by microphages. But macrophages participate in chronic inflammation as seen in tuberculosis and some fungal infections.

The main steps of phagocytosis are :

(i) Recognition and attachment of microbes as foreign particles to leucocytes.

(ii) Engulfment of the microbes by leucocytes and formation of a phagocytic vacuole in the cytosols.

(iii) Intracellular digestion and killing or degradation of the ingested particles.

Microbes are first coated by opsonins (serum proteins) or immune opsonins). These opsonins bind to receptors on the leucocytes. Then, phagocytosis of the opsonised bacteria is noticed.

The main opsonins are :

(i) Fc fragment of immunoglobulin IgG.

The corresponding receptors on the leucoytes is FcyR which recognizes Fc fragment of IgG.

(ii) C3b, (an opsonic fragment of C3) is elaborated by activated complement.

(iii) Collectins (carbohydrate-binding proteins). The collectins bind to bacterial walls.

Binding of the opsonised microbes to FcyR (receptors on the leucocytes) is followed by its engulfment by leucocytes. The engulfed organisms are killed and degraded by phagocytes. The reactive oxygen metabolites destroy the ingested bacteria and oxidized halide HOCL (arising from conversion of H_2O_2 to hypochlorous acid) in the presence of halide like Cl. With the help of enzyme myeloperoxidase (MPO, a powerful antimicrobial agent), lysomal hydrolase degrades the killed bacteria. MPO deficient cells also destroy the bacteria with the help of superoxide (O2−), hydroxl radical (HO)· and hydrogen peroxide H_2O_2.

Phagocytes kill the bacteria by using oxidative mechanisms in the cells which show respiratory burst of superoxide products during phagocytosis. The use of extraoxygen to elaborate reactive oxygen metabolites which destroy the engulf bacteria. The mechanism of respiratory burst is activated by membrane bound oxidase i.e. NADPH. This is marked by reduction of two molecules of molecular oxygen ($2O_2$) into two molecules of superoxides ($2O_2^-$). Activated neutrophiles show rapid burst of superoxide radicals.

Products of macrophages

Lysosomal granules in these cells are released into phagocytic vacuoles. Fusion of these vacuoles (phagosomes with lysosomal granules result in formation of phagolysosomes. Pathogenic bacteria are destroyed by proteases and other enzymes present in the lysosomal granules. Engulfment of the pathogenic microbes involves binding the opsonised bacteria to the FcyR (a corresponding receptor on the leucocytes). Pseudopods of the leucocytes e.g. neutophiles surround the foreign opsonised particles and enclose them within the phagosomes (phagocytic vacuoles) created in the cell cytosol. The limiting membrane of the lysosomal granules fuses with the membrane of phagosomes to form a structure called phagolysosome.

The main enzymes in the lysosomal granules associated with mediation of inflammantory reactions and killing of bacteria are as follows :

(i) Acid hydrolases.

(ii) Acid phosphatases.

(iii) Proteases.

Acid proteases degrade bacteria and cellular debris in the phagolysosomes.

(iv) Estrases like phospholipase B and lysophospholipase.

(v) Nucleases like ribonucleases and doxyribonuclease.

(vi) Oxidoreductase myeloperioxidase and histaminases.

Repair and regeneration

Responses to damage in the bodies of organisms are known as reactions of the tissues. These reactions include regeneration which occurs to compensate the loss of some structures with similar kind of tissues. In animals, there may be destruction of a structure by chemicals and trauma etc. and in such case, the accumulated materials like blood, fluid, destroyed bacteria and dead tissues are lysed and removed and processes of these kinds are called reparatory processes. The gap in the tissues due to destruction is filled up with newly formed tissues. Injurious factors i.e. mechanical, bacterial and chemical factors etc. cause destruction of both parenchymal cells and stromal structures. Repair by regeneration is shown only by some parenchymal cells in organs e.g. livers cells or epidermal tissues. But in other organs, the repair is affected by replacement of non-regnerated cells or tissues like connective tissue i.e. the processes of granulation tissues formation by angiogenesis and fibrosis and scarring. Epitheliasation of the defect filled by granulation tissue results from migration of proliferative epithelial cells from the margins of the wounds. And the main features of the processes i.e. the different intention are given in the Table 10. Injuries cause destruction and loss of tissues. Wound space or defect is formed in the injured tissues. This extravascular space gap is filled with blood escaping from the vessels in the zone of injuries. In such events, platelets show the earliest step of adhesion and aggregation at the injured site to close the cut in the vessels to prevent further loss of blood. Platelets are activated by contact with collagen in the extracellular matrix (ECM).

The main steps in formation of granulation tissues for filling the defect in the injured tissues are as follows.

1. Angiogenesis (formation of new blood vessels e.g. capillaries).

2. Migration and proliferation of fibroblasts.
3. ECM (extracellular material) deposition.
4. Remodeling (maturation and organization of the connective tissue) These processes give maximum strength to scar.

Repair is effected by a specialized type of tissues called granulation tissue which contains several newly formed blood vessels.

Some main information about granulation tissue

Granulation tissue reveals pink soft and granular appearance on the surface of wounds. Histologicaly, the main observations on the granulation tissues include newly formed blood vessels i.e. capillary loops or buds by a process known as angiogenesis. Then, the next step is that of fibroplasia (fibrosis) in which proliferation of fibroblast is observed in the granulation tissues (a framework of newly formed capillaries aided by fibroblasts). Fibroblasts grow and multiply around these blood vessels with this arrangement alongside and parallel to these blood vessels. In this processes of fibrosis, one sees another phenomena, namely: (1) emigration of leucocytes like neutophiles and macrophages, (2) proliferation of fibroblasts in the damaged tissue to make deposits of ECM formed by growing fibroblasts.

The newly formed blood vessels are leaky allowing escape of plasma proteins and red cells in the extravascular spaces. Increased vascular permeability is followed by escape of plasma proteins like fibrinogen and plasma fibronectin in the extracellular matrix to form a provisional scaffold for ingrowth of fibroblasts and angioblasts (precursors of endothelial cells of the blood vessels). Macrophages form cellular components of granulation tissues, degrade and remove extracellular material, fibrin and debris etc. from the injured site to complete wound healing.

Table-9 Important roles of some growth factors

Cellular products or growth factors	Roles
PDGF	Monocytes chemotaxis, migration and proliferation of fibroblasts.
TGF-β	As above and increased synthesis of collagen and fibronectin
Metalloproteinases e.g. collagenase, gelatinase and stromelysins	Degradation of ECM
VEGF	Angiogenesis
PDGF, TNF and TGF β	Collagenase synthesis.

EGF, FGF, TNF and PDGF

Such reparative steps occur quite early in inflammation after injuries in tissues and may be noticed in 24 hours after damage to tissue by non infectious factors. By three to five hours post-injury one notices proliferative changes in fibroblasts and vascular endothelial cells. Contact inbihition controls the proliferation of cells engaged in repair through intractions with adjacent cells. Activated wound microphages because of hypoxia, and acidity (increased in concentration of lactic acid) elaborate growth factors (Table-9 and 11) to initiatie the fibroblastic phase of wound healing. The cavity in the wound is filled with a provisional extravascular gel or matrix in the initial stages of repair. The newly formed matrix acts like a substrate for subsequent leucocytes migration e.g. migration of neutrophiles followed by movement of macrophages (Table 10). Activated platelets due to contact with fibrillar collagen (a component of ECM), elaborate material like fibronogen, fibronectin and thrombospondin etc. Extracellular matrix molecules, aggregated platelets in the injured tissues and fibrin enrich the extracellular matrix (ECM) or extravascular gel for the reparative and defensive roles of different cells like angioblasts, fibroblasts and leucocytes e.g. microphages and macrophages etc.

Transcriptional factors in healing of injuries

Healing (repair) of wounds or injuries in the tissues due to cell death and mechanical damage etc. is achieved by cell proliferation i.e. regenerative activity and cell replication governed by certain chemical factors. One sees an uncontrolled growth (cell proliferation) due to an excessive role of stimulators or deficient inhibitors in the malignant neoplasm e.g. cancerous growth. Fibroblasts, endothelial cells, chondrocytes, osteocytes and smooth muscles cells are subject to influences of autocrine, paracrine and endocrine signaling and proliferate following injuries in tissues and organs as per need of the bodies in organisms. In short, the growth of tissues is altruistic. For example, healing process of the gap in the wound ceases with the filling of the gap with granulation tissue and completion of the epitheliasation. These cells are stable cells remaining quiescent in the adults. Neurons, skeletal and cardiac muscle cells are permanent cells failing to undergo mitosis in the postnatal life.

Cell growth in the reparative process is marked by the following steps.

1. Binding of the signaling agent (a molecule) called ligand to a specific cell surface receptor. It is a ligand- receptor reaction on the cell surfaces. Growth factors like epidermal growth factor (EGF), fibroblast growth factor (FGF) and platelet derived growth factor (PDGF) bind to their specific cell surface receptors.
2. Signal transduction activity to nuclei of the cells by certain pathways, namely (i.) Phosphoinositide-3- kinase (PI-3 - kinase) pathway. (ii) Inositol-lipid (IP3) pathway (iii) mitogene activated protein kinase (MPP kinase) pathway (iv) Cyclic adenosine monophosphate (CAMP) pathway and (V) JAK/STAT pathway.

The effects produced by these pathways are:

Pathway	Effects
1. PI-3 Kinase pathway	Cellular survival through activated cellular kinases.
2. MAP kinase pathway	Entry of the quiescent cells into the growth cycles through activated phosphorylated proteins. MAP kinase (translocated in the nuclei) phosphorylates specific factors like c-jun and c-fos causing activation of gene expression and uncontrolled cellular proliferation as seen in neoplasia e.g. cancer.
3. IP3 pathway	Related to cell growth and metabolism through activated protein kinase and subsequent phosphorylation of cellular components.
4. CAMP pathway	Expression of target genes through activated protein kinase.
5. JK/STAT pathway	Mediating functional responses.

The different five signal transduction systems involved in the transcription processes pass on information to the nuclei of cells in which specific changes regulate the gene expression. Regulatory factor (also called transcription factors) are active at the level of transcription of genes. Thus transcription plays important role in the normal cell growth by initiating DNA transcription in the repair of wounds

Two important steps in the angiogenesis in healing of wounds are :

1. **Vasculogenesis**

It refers to formation of a vascular network by proliferation of precursors of endothelial cells called angioblasts.

2. Angiogenesis (neovascularisation).

3. It refers to formation of newly formed vessels by preexisting vessels which give rise to buds or sprouts. It is a very important phenomenon in repair, fibrosis and chronic

inflammation, neoplasia and formation of collateral circulation etc.

The main events in formation of new blood vessels are as follows:

1. Proteolytic degradation of the basal membrane (BM) of the parent blood vessels at the wound site for formation of capillary sprouts and cell migration e.g. angioblasts.
2. Migration of endothelial cells towards the angiogenic stimuli and their proliferation.
3. Maturation of endothelial cells and there remodeling into capillary tubes.
4. Recruitment of pericytes (periendothelial cells) and vascular smooth muscles cells for providing support to growing or proliferative endothelial cells.

All these aforesaid processes are controlled by interaction between vascular cells, several growth factors and extracellulalr matrix.

Fibrosis (Fibroplasia)

It is a supportive or strengthening step after the process of angiogenesis (marked by formation of new blood vessels) and deposits of loose extracellular material at the site of wound healing. This process of fibrosis reveals emigration and proliferation of fibroblasts and deposition of extracellular matrial in the zone of wound healing. In early stages, granulation tissues reveal numerous blood vessels, loose extracellular material, and oedematous appearance. Vascular endothelial growth factors (VEGF) promote angiogenesis and cause marked vascular permeability. Plasma proteins like fibrinogen and fibronectin are deposited in ECM to give a provisional stromal support to fibroblasts and angioblasts. Interlukin-1, TNF, PDGF, EGF, FGF and TGF α cause migration and proliferation of fibroblasts at the injured sites. Macrophages clear cellular debris, fibrin and other foreign substances. The factor TGF- α causes migration and deposition of collagen and fibronectin. It also shows chemotaxis towards monocytes and triggers angiogenesis at the wound sites.

Tissues remodeling

Tissues remodeling which has the features of the rebuilding, strengthing and change in the constituents of extracellular material is noticed in wound healing and chronic inflammantory reactions. Remodeling refers to a net result of synthesis of extracellular material and degradation at the injured site under repair. The extracellular matrix in the wound space or gap consists of proteins derived from platelets and plasma cells. Macrophages migrate into the ECM of wound space. Fibroblasts deposit fibronectin in the healing wound and it acts as substrate for cell attachment. The components of extracellular matrix consist of hyaluronic acid and fibronectin permitting micromovement of cells in the early stages of repair. Decrease in concentration of hyaluronic acid and increasing proteoglycans favour cell attachments and a lack in the movement of cells. Then, the cells in the ECM of the wounds mature and differentiate into endothelial cells. Endothelial cells form capillaries and fiboblasts proliferate to form collagen and deposits of fibronectin. Oxygen favours synthesis of collagen. Fibroblasts differentiate in myofibroblasts in the granulation tissue. The last step in the wound healing is maturation and remodeling of the extracellular matrix for providing tensile strength to the scar. Granulation tissue in the wound space is replaced by a scar due to changes in the composition of the extracellular matrix. Growth factors degrade extracellular matrix. Metalloproteinases degrade collagen and ECM components and play important role in remodeling. In tissue remodeling for formation of scar, adequate production, digestion, aggregation, and orientation of collagen fibrils and fibers in the direction of the forces acting on the tissue are important processes till a balance is obtained between production and break down of collagen. All the changes are intended for enhancing tensile strength in a scar. Collagen deposition and remodeling of the collagen fibrils gives strength to fibers of connective tissue. One notices large collagen bundles with cross linking between collagen fibrils. The collagen deposited in the fibronectin in a haphazard manner is digested by the enzyme collagenase. Macrophages, fibroblasts and endothelial cells form collagen in the granulation tissues. Growth factors influence production of ECM by fibroblast and excessive and exuberant mass of connective tissue in a skin wound is called keloid (a tissue mass rich in collagen). The last step noticed in the healing of a wound is an avascular scar rich in dense connective tissue and collagen with a tensile strength and intact covering of epidermis. Scarring occurs in the wound within a month or so.

Healing or repair of defects in the tissues (Table 10 and 10a) can take place by certain processes, namely.

1. Healing by primary intention, and
2. healing by secondary intention.

Healing by primary or 1st intention

Healing by primary intention has the following main steps.

1. Cessation of haemorrhages.
2. Cohesion of the edges of the wound.
3. Filling of the cleft by fibrin.

The edges of the wound become hyperaemic and from these congested vessels, neutophiles and lymphocytes migrate and move into the extracellular matrix of the wound itself. Leucocytes migration occurs during the first 2 hours. Later on, histiocytes and advential cells migrate into the wounds and these cells phagocytise the oozed out blood, damaged or dead tissue cells etc. Proliferations of cells begin by 2 hours and after 48 hours, miotic figures are seen in the connective tissue and endothelial cells etc. Young or fusiform connective tissue cells are seen at the edges of the wound and these grow into the wound. These, later, unite with the capillary buds and connective tissue cells from the other sides of the wound to establish continuity. Later, the young connective tissue cells form fibres and finally a scar is formed after decrease in cellularity and vascularity in the young tissue. At first, the scar is red and later, it becomes pale and consists of an abundance of collagen (fibres) and relaltively few cells. Papillae are not usually found in the epithelium over the wound. Epithelium grows out from the edges of the wound due to mitotic activity. After 4 to 6 weeks, elastic fibres usually appear, pigmentation is partially present in the scar or may even be absent, Healing by 1st intention takes place in about a week and the following are the required conditions for such type of healing.

1. Wound should be aseptic.
2. Damage should not be extensive.
3. The edges of the wound should be in apposition.

Only evidence of injury after repair is a fine white line of scar tissue.

Healing by 2nd intention

This process is applicable in case of healing of the open wound which may be created by destruction of a large amount of tissues. Such wounds usually get infected and show inflammatory changes. Repair of the lost tissue is attained through formation of granulation tissue.

Gross appearance

Granulation tissue arises from formation of the vascular connective tissue which looks red and granular to the naked eye. Minute projections (or buds) are formed by budding capillaries originating from the endothelium of the preexisting capillaries in the inflamed reparative zone.

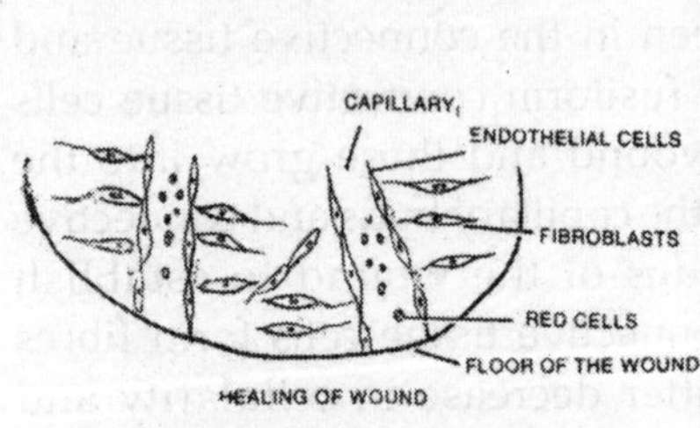

Fig. 28 Healing of wound. Note the vessels and fibroblasts at right angles to vessels.

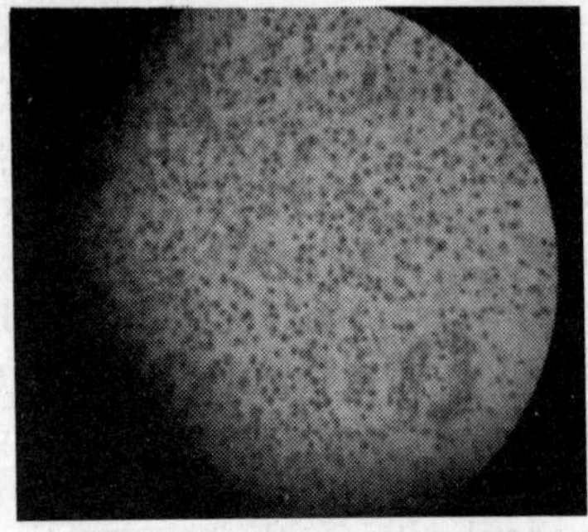

Fig. 29 Young granulation tissue, skin. Note the newly formed blood vessels and round or oval cells (fibroblasts or fibroblasts cut transversely) H & E x 400.

Microscopic appearance

The following are the main histological features :

1. Swollen up capillaries in the floor of wound project into the direction of the wound and the buds i.e. proliferated endothelial precursory angioblasts) are formed by them. Angioblasts proliferate, mature and change into capillary tubes.
2. Anastomosis of the neighbouring capillary processes to form capillary loops. These loops elongate and form further loops.
3. Arrangement of the fibroblasts mainly in a parallel way with the newly formed blood vessels (neocapillaries) in the intermediate layer or zone arising through angiogenesis . Such fibroblasts are spindle shaped cells with faintly staining nuclei (Fig. 28)
4. Arrangement of fibroblasts at right angles to the vessels in the deeper layer (Fig. 29).
5. Continuation of the proliferation of fibroblasts and formation of neocapillaries till the filling up of the defect or gap.
6. Gradual removal of fibrin, leucocytes and necrotic tissue.

At later stages, the new tissue becomes less cellular and fibroblasts become spindle shaped which form collagenous fibrils.

The gap is occupied by the fibrous tissue which may contract to obliterate the blood vessels. The damage is replaced by new vascular connective tissue which is covered up by the growing sheets of the epithelium from the periphery or margins of the wound.

Lastly, fibrous tissue contracts and newly formed vessels are obliterated to form what is known as scar. As it is very difficult to separate chronic inflammatory changes from reparative changes i.e. formation of the granulation tissue, inflammatory cells like neutrophiles, lymphocytes, plasma cells, mononuclear cells or macrophages are seen in both types of these pathologic processes or changes. Papillae, hair glands or pigments are not found in the covering epithelium over the granulation tissue.

Healing by 3rd intention

Granulating surfaces unite under complete asepsis and the walls of the wound are also in perfect affrontment but it rarely happens.

Table- 10. Some main features of healing (repair) of wounds

Types of Healing	Main features
1. Healing by first intension For this, important condition is an aseptic wound with its edges in apposition	1. Absence of inflammatory or infectious processes in the injured tissues. Completion of wound healing occurs in about a week. (i) Presence of clotted blood containing blood constituents and fibrin and a surface clot (called scab) over the wound surface. (ii) Migration of neutophiles into fibrin clots within 24 hrs. thickening of the epidermal cells at the cut adges of the wounds and growth of the epithelial cells from the edges of the wounds. B.M. (Basement membrane) material (non fibrillar collagen, laminin, heparin sulphate, proteoglycans and some glycoproteins) are deposited with the fusion of the epithelial cells under the scab to form an epithelial covering. (iii) Neutrophiles replaced by macrophages and occupation of the incisional space by collagen fibres i.e. so called generation tissue with thickening of the superficial epidermal covering 2. Neovascularization

	(i) (angiogenesis) and newly formed granulation tissue filling the incisional space. Epithelial cells covering the wound filled with granulation tissue mature and undergo by 5th day keratinisation. (ii) Proliferation of fibroblasts, increasing collagen fibers and absence of leucocytic infiltration increased vascularity and edema are main changes during the second stage of wound healing. (iii) Presence of a scar composed of connective tissue and superficial covering of intact keratinised epidermal cells, no presence of dermal or epidermal appendaces dedacted in the scaring of incisional injuriy. At last, avascular and a dense connective tissue scar with maximum strength and collagenation formation is noticed within a few months after wound healing.
2. Healing by second intension It is noticed in the infected open wounds with extensive tissue damage and separated edges. Such wounds in the tissues are called open wounds. Healing of infarct, abscess and surfaced wounds are examples of healing by second intention	(i) Large defects in tissues contaning infected necrotic debris and exduate rich in the inflammatory cells e.g. neutrophiles etc. Presence of inflammatory reaction in the open wounds. (ii) Filling of the defects in the injured tissues by extensive amount of granulation tissue (iii) Occurence of wound contraction with in the tissues affecting mobility of organs e.g. cardiac movements in chronic pericarditis or intestinal movement in chronic intestinal wounds
3. Healing by third intension	i. It is marked by union of the granulating surfaces under complete asepsis and perfect affrontment of the walls of the wounds. But, a healing of this kind rarely occurs

Note: Both inflammatory and reparative processes in the injured tissues are concurrent processes and the proliferation of both angioblasts and fibroblasts runs side by side in concert in the injured or damaged tissues. Proliferation of angioblasts constituting newly formed blood vessels i.e. granulation tissue in wound healing is usually followed by multiplication of fibroblasts in the granulating mass of capillaries. Scarring is complete in a month or so and a connective tissue formation rich in collagen with no inflammatory exudate and lack of vascularity with intact covering of epithelium over the wound surface is noticed.	

Table:-10a. Main components of ECM (extracellular matrix) and their important functions in wound healing

Components	Sourceses/products of some activated cells	Roles/Functions
Fibronectin	Platelets, microphages, endothelial cells, fibroblasts and epithelial cells.	1. Provision of substratum or scaffold for cells like fibroblasts to adhere, proliferate and migrate. It also acts like scaffold for deposit of new ECM (e.g. collagen). ii. Degradition of ECM in wound healing invasion and metastasis of maliginant neoplasm. Fibrin, a component of ECM binds to cells via specific receptors. iii. Chemotactic for macrophages, stimulation of endothelial cells and fibroblast for the sake of movement of migration iv. Ligand for integrins (adhesion molecules) on cells and opsonins for phagocytes. In other wounds, it binds to members of the integrin receptors.
2. Collagen (Nine types of collagen detected) Fibrillar collagen includes types i, ii and iii . Types iv, v and vi are non fibrillar of amorphous collagen in interstitial tissue and basement membranes (BMs). Collagen synthesis occurs in ribosomes of endoplasmic reticulum.	Fibroblast	Provides tensile strength to healing wounds and acts like major. components of ECM frame work.
3. Elastin It occurs in uterus, skin, ligaments and blood vessels e.g. aorta etc.	Fibroblasts	i. Gives elasticity or ability to recoil to ECM of tissues like uterus, lung, skin and blood vessels etc.
4. Proteoglycans, adhesive glycoprotein and integrins. Heparin sulphate, chondroitin substratum sulphate and dermatan sulphate are examples of proteoglycans.	Fibroblast	i. Links or binds ECM components two one another and also to cells and act like adhesive proteins, causes and strengthens cell attachment to ECM. ii. ECM components i.e. collagen fibrin and proteoglycans, bind to fibro nectin

5. Laminin (glycoprotein). It occurs in the basement membrane of cells.	Epithelial cells, monocytes and endothelial cells	i.Mediates cell attachments to connective tissues substrate ii.Causes attachment of cells to basement membrane, influences adherence of cells to collagen i.e. type iv. iii. Binds to members of the integrin receptors family iv. Certain components in the ECM e.g. fibronectin and laminin bind to integrin.
6. Hyaluronic acid (glycosaminoglycan polysaccharide)	Fibroblasts	i. Helpful in cell movement due to weakening caused to cell attachment with the substratum. ii. Facilitates adhesions of cells and substrate i.e. scaffold.

Table-11. Growth factors, their sources and roles in the wound healing

Growth factors	Sources / Products of activated cells	Roles / Functions
1.PDGF (Platelet derived growth factor)	Platelates, smooth muscle cells, microphages and endothelial cells	1.Responsible for migration and proliferation of fibroblast, smooth muscle cells and monocytes to i. Show chemotactic action for fibroblast smooth muscle cells, neutrophiles and monocytes . ii. Mitogenic influences on fibroblasts induces increased fibroblast production i.e. more formation of collagen and elastin in the E.C.M. (extra cellular matrix)
1. EGF (Epidermal growth factors)	Microphages and platelets	2. Mitogenic affects of epithelium and fibroblast in vitro. Induces proliferation of hepatocytes and precocious tooth eruption i. Responsible for proliferation of endothelial cells, epithelium and fibroblast. ii. Reduces synthesis of collagen by fibroblast.

3. TGFβ (transforming growth factor β)	Macrophages, T-cells and platelates	3. Induces synthesis and secretion of PDGF. Mitogenic effect on cells e.g. fibroblasts. i. Shows chemotaxis for fibroblasts and leucocytes. ii. Causes production of ECM i.e. components of ECM like collagen and fibronectin etc.
4. FGFs (fibroblast growth factors) These growth factors are divided into (1) acidic FGF and (2) basic FGF .	Different kinds of cells	i.Induces formation of new blood vessels (angiogenesis) ii. Causes migration of macrophages, fibroblasts and endothelial cells in the scaffold (ECM) in injured or damaged tissues.
		iii. Responsible for lung maturation and development of skeletal muscle. iv. Induces hematopoiesis and mitogenic effect for endothelium, fibroblasts-chondrocysts and smooth muscles cells etc.
5. IL-1 cytokine	Activated macrophages and endothelial cells	i. Shows chemotaxis for leucocytes and epithelial cells ii Induces proliferation of fibroblasts, smooth muscle cells and keratinocytes. iii. Induces production of ECM components i.e. collagen and elastin.
6. TNFα (A cytokine)	Activated macrophages	i. Mitogenic effect on fibroblasts. ii. Shows angiogenic effect in the wound healing and enhances production of collagen. iii. Causes reabsorbtion of bone or cartilage
7. Interferon γ (A cytokine)	Activated lymphocytes	1. Decreases collagen synthesis and fibroblast proliferation.

Repair of fracture

Periosteum and surrounding tissues are injured in cases of fractures of bones. Fracture in the bone causes haemorrhages and in case, the fracture is aseptic and there is not much destruction of tissue or excessive displacement of the fractured ends, repair occurs without much delay. Blood clot stimulates neighboring connective tissue cells to proliferate.

The blood vessel in periosteum, and medulla dilate and lymph exudes with emigration of leucocytes which remove the bits of blood clot. Later, proliferative changes of cells take place in periosteum, medulla and the Haversian system of the bone with formation of cells and new blood vessels. These cells and new vessels invade the area occupied by the blood clots. The newly formed granulation tissue consists of osteoblasts and young blood vessels and the newly formed granulation tissue encloses the fractured ends with the formation of a spindle shaped mass. Lime salts are also deposited in the new tissue enclosing the fractured ends with the formation of a spindle shaped mass. Lime salts are also deposited in the new tissue and a spongy mass of bone formed in this way is called provisional callus. The provisional callus acts like a scaffold and keeps the fractured ends in apposition, It is divided into 3 parts :

1. External callus (outer most part lying under the periosteum)
2. Internal callus (callus in the medullary cavity of the bone)
3. Intermediate callus (callus between the fractured ends).

Osteoblasts form layers of compact bone with Haversian systems in the line of support given to the provisional callus. Oesteoclasts which are multi nucleated cells remove the external callus. Such processes ultimately lead to the formation of a definitive callus between fractured ends of the bones.

Both ends are united by compact bone and the process of healing is, thus, finally completed. Granulation tissue can also be found in an organ, at the edges of necrosis and the infracts during the replacement of thrombi or blood clots or exudates and in chronic inflammation. Pyogenic membrane in the walls of fistulae and abscesses is also in nature of the granulation tissue. Reparative processes are basically following the same principle in both osseous and non-osseous tissues but the osteoblasts in place of fibroblasts play the role to unite broken bones in osseous tissues.

Regenerative changes

Regeneration refers to the replacement of lost material, physiological regeneration (i.e. replacement of the lost tissues by a smilar kinds of tissues) is seen in skin, mucous membranes, glands, hair, haematopoietic tissues and sexual organs etc. In pathological regeneration, a new tissue replaces the defects produced by diseases. After severe damage, there is no complete regeneration (i.e. formation of a replica of the original tissue). When regenerated tissue differs from the original, it is called atypical regeneration. Regeneration is to greater extent in embryonal parenchyma than in the specialized parenchyma. Neurons are not capable of regeneration. The blood vessels and connective tissue cells are capable of extensive regeneration in the bodies of higher animals.

Regeneration of epithelium

Epithelium of the stratified squamous epithelium regenerates from stratum germinativum. The cornea regenerates rapidly and intestinal epithelium regenerates from the crypts within the walls of intestine. Skin cells have been induced into pluripotent state with features of embryonic stem cells.

Fibrillary connective tissue

Defects or damage in the fibrillary connective tissues are replaced by granulation tissue originating from the pre-existing connective tissue cells of the injured region. The young fibroblasts are large, fusiform or polygonal cells. Fibroblasts also form elastin.

Adipose tissue

An accumulation of fat droplets in large, round and polyhedral cells with abundant cytoplasm gives rise to fat cells or adipose tissue. The adipose tissue also arises from nucleus containing cytoplasmic remnants of old fat cells. Accumulating fat in the cell pushes the nuclei to one side.

Cartilage

Cartilage originates from chondroblasts in the perichondrium. At first, the newly formed tissue from perichondrium is fibrillar which later, undergoes homogenization of the protoplasmic portion to form hyaline of the cartilaginous ground substances.

Osseous tissue

Regeneration of the osseous tissue involves the periosteum and endosteum. Bone is also formed from perichondrium. Osteoblasts first appear and then bone is formed through proliferation and development of the exoplasma. The bone cells are united with one another by fine processes to form a woven bone. Oseoblasts arranged in layers are separated by exoplasm and these give rise to lamellar bone.

Blood and lymph vessels

Endothelial cells of existing capillaries swell and divide mitotically to form blood channels. Lining cells of the newly formed vessels have plump endothelial cells. The endothelial cells form solid buds, later, become canalized to form capillary tubes. The connective tissue cells and smooth muscle cells strengthen the endothelial cells and also form fibrils to strengthen the walls of the blood vessels. Arteries and veins also arise from such processes.

Regeneration of haematopoietic tissue

The lymphocytes regenerate in the spleen, lymph nodes and lymphatic tissue in the mucous membranes. The lymphocytes in connective tissue are obtained from the blood or lymph vessels. The neutrophiles, eosinophiles and basophiles are formed in bone marrow (i.e. from the spleen, liver and bone marrow).

Muscle tissue

Smooth muscle cells have a slight regenerative capacity and smooth muscle cells undergo hypertrophy as seen in the uterus during pregnancy or in gastric and intestinal wall in cases of stenosis. Striated muscle cells have greater capacity for hypertrophy than hyperplasia. Cytoplasmic buds are formed from the existing fibres to regenerate the skeletal muscle cells.

Nervous tissues

The neuroglial tissues show regeneration. The stumps of nerve fibres in the central nervous system may regenerate after being sectioned into parts. The axon cylinders are formed first and then the myelin sheaths also regenerate. Neurons do not regenerate in the brain, spinal cord or in the peripheral ganglia. A marked regeneration is noticed in the neuroglia (i.e. glial cells or astrocytes etc.) in the brain.

Peripheral nerves

Damage to the fibres causes degeneration in the proximal stump of the fibres up to the nearest Ranvier's node. The disal stump regenerates up to the ramifications. When the degenerative changes have been completed, regeneration commences in both distal and proximal stumps due to mitosis in the Schwann cells. The myelin sheaths are formed at a later stage. Large cells, called neuroblasts are also formed and at a later stage, these large cells, called neuroblasts, later, give, rise to band like structures which differentiate into neurofibrils. Connective tissue is formed from epi and endoneurium and newly formed fibres course through it. Large defects give rise to connective tissue scars. Regenerated nerve fibres get woven with endo and perineural connective tissue to form a tumour like swelling called amputation neuroma i.e. a mass of granulation tissue at the site of amputation. It is also believed that the nerves do not regenerate in the brain and spinal cord. The defect in the nervous tissue is occupied by neuroglial proliferation. The axon of the proximal stump grows into the distal severed portion of the nerve and exerts a chemotactic effect on the new growing axon. If the severed ends of nerves are kept in apposition, the new fibres enter into the old sheath.

Chapter **7**

Diseases of Immunity/ Immunopathology

Hyperactive immune system in the body causes death of organisms. An allergic reaction to the sting of a bee may be fatal for an individual. A deranged immune system fails to differentiate self from non-self in the body and autoimmunity (immunity against one's tissues) may develop in the body. Autoimmune disease may arise in the body of individuals. Amyloidosis which is also considered to be disease associated with immune system is marked by formation of abnormal proteins which are fragments of immunoglobulins and in some cases; these are deposited extracellularly in the tissues. The pathological condition produced in animals is called amyloydosis (amyloid infiltration). Humoral immunity is caused by antibodies (glycoproteins) which are produced in the body against antigens or immunogens (substances mostly foreign to a host).

Disorders of the immune system are produced by disturbances of immune system which are marked by hypersensitivity reactions. These hypersensitivity reactions are mechanisms of immunologic tissue injuries in the body of an animal.

The diseases or disorders of immune system are:

1. Hypersensitivity reactions

These are marked by immunologic injuries in cells or tissues of the body.

2. Autoimmune diseases

The pathogenesis of the diseases lies in the immune reactions against self.

3. Immunologic deficiencies syndromes these conditions arise from some defects in immune response.
4. Diseases or disorders associated with some immunologic response e.g. amyloidosis in individuals or animals.

Hypersensitivity reactions

Exogenous factors e.g. drugs, pollens, chemicals and transfusions reaction following intravenous administration of blood produce hypersensitivity in the body immunologic mechanisms mediating the tissues injuries or disease processes are the bases of classifying the hypersensitivity diseases.

Cytotoxic hypersensitivity arises from antigen- antibody reactions on the surface of the cells. These reactions activate complement system causing cell lysis or cell destruction.

The following are the important types of hypersensitivity diseases.

1. Type I disease (anaphylactic type) is marked by formation of IgE antibodies, release of vaso- active peptides and spasmodic substances acting on vessels and smooth muscles and some mediators are also released from basophiles and mast cells etc.

2. Type II disorders (cytotoxic types)

These are marked by formation of IgG and IgM etc. and binding of humoral antibodies on target cells which are injured, phagocytosed or may be lysed by activated complement. In short, there is an antibody dependent cellular cytotoxicity (ADCC) in this cytotoxic disorder.

3. Type III disorders (immune complex diseases)

These are disorders marked by humoral antibodies to antigens and activation of the compliments. Activated compliments attract neutrophiles which release lysosomal enzymes and other toxic fraction. Lysosomal enzymes and toxic free radicals (reactive oxygen species) produce tissue injuries. Arthus reaction, serum sickness and some kinds of acute glomerulonephritis are examples of immune complex diseases.

4. Type IV disorders (Cell mediated or delayed hypersensitivity)

These are marked by presence of sensitized T lymphocytes. T lymphocytes release lymphokines and T cell- mediated cytotoxicity develops in the body. In short, there is a cell mediated immune response. Cell mediated cytotoxicity is seen in tuberculosis, transplant rejection and contact dermatitis.

Type I hypersensitivity (Anaphylactic type)

This anaphylactic type of hypersensitivity develops in the sensitized individuals within a few minutes. The antigens bind to specific antibodies on mast cells or basophiles in previously sensitized individuals.

The important features of this disorder are:

(1) Initial response seen as vasodilatation and vascular leakage.

(2) Late phase response.

It is noticed in 2 to 8 hours and marked by infiltration of tissues with eosinophiles, neutrophiles, monocytes, basophils and CD+T cells causing tissue destruction.

(3) Release of primary mediators as amines (histamine), enzymes (proteases and acid hydrolases) and proteoglycans.

(4) Release of secondary mediators like cytokines and lipid mediators (arachidonic acid metabolites).

Type II hypersensitivity (Cytotoxic type)

It is marked by either complement dependent reaction or antibody mediated injury to target cells. The antibodies produced are directed towards antigens present on the surface of the target cells or other tissue components. In short, type II hypersensitivity is marked by binding of the antibodies to normal or altered cell surface antigens of target cells. Antibody dependent cell mediated cytotoxicity (ADCC) is marked by coating of target cells by antibodies like IgG antibodies. The cells coated with antibodies are killed by cells like non -

sensitized neutrophiles, monocytes and NK cells etc. These killer cells (NK cells) have Fc receptors for IgG. ADCC is also a mechanism seen in tumour immunity working in concert with cellular immune response. There is no fixation of complement in the antibody dependent cell immediated cytotoxicity. Instead of this, only leucocytes destroy the tumour or other cells. Cell lysis without phagocytosis is a feature of ADCC. Only Fc fragment of IgG antibodies coating the tumour cells and Fc receptors on non- sensitized leucocytes play important role in killing of tumour cells in ADCC.

In complement dependent reactions, there is either lysis or phagocytosis of the target cells. Direct lysis of the target cells is caused by reaction of antibody (e.g. IgM or IgG) with specific antigens present on the surface of the cells. This antigen-antibody reaction activates complement system resulting in the formation of membrane attack complex (C5-9). This membrane attack complex drills holes through the cell membrane with ultimate lysis of the target cells. Transfusion reactions involving red cells, platelets and white cells because of administration of incompatible blood from a donor into a recipient and erythroblastosis foetalis because of antigenic differences between the mother and foetus casuses haemolysis arising from reaction between mother's antibody and red cells of foetus are examples of type II hypersensitivity.

Another way destruction of target cells is through pahagocytosis when there is no fixation of the complement in the Ab-Ag complex. Antibody C3b (a complement) is fixed to the cell surface of the target cell through the process of opsonisation. The opsonised cells like blood cells in complement dependent reactions are either lysed or rendered susceptible to phagocytosis.

III. Type III hypersensitivity (also called immune complex disease)

It is marked by formation of antigen-antibody complexes (Ab-Ag complexes) which activate the complement. Neutrophiles attracted to the deposits of immune complex, release lysosomal enzymes and toxic intermediates causing destruction of the target cells. Exogenous factors (e.g bacteria, viruses and fungi etc.) and endogenous factors (e.g. nuclear antigens, tumour antigens and immunoglobulins) form immune complexes with these specific antibodies. These complexes are deposited in organs like kidneys or joints and cardiac valves.

Glomerulonephritis, valvular endocarditis are examples of type III hypersensitivity due to immune complexes. Immune complex deposits in the tissues cause immune complex mediated inflammation, release of neutrophiles, lysosomal enzymes and fibrinoid necrosis in the walls of blood vessels.

Immune complexes are formed in blood circulation or extravascular sites. Acute serum sickness arising from administration of foreign serum e.g horse antitetanus serum is an example of systemic immune complex disease. Immune complex deposits in tissues e.g. joints or kidneys and extravascular areas cause activation of the complement system, production of chemotactic factors (e.g. C5 fragment) and release of C3a and C5a (anaphylatoxins). The effects of complement fixation are increased vascular permeability, oedema, cytolysis, necrosis and formation of membrane attack complex (C5-9) . Arthus reaction is an example of local immune complex disease.

Type IV hypersensitivity (Cell mediated hypersensitivity)

It is marked by hypersensitivity reaction due to specifically sensitized leucocytes e.g. CD4 T lymphocytes and CD8-T cells. Delayed type hypersensitivity seen in tuberculosis is an example of type IV hypersensitivity. This is also produced by parasites, protozoa and fungi.

The active component of tuberculin is a protein-lipoploysaccharide. Intracutanuos or intradermal injection of tuberculin is followed by formation of hot , painful and indurated swelling at the site of injection in 8 to 12 hours and the swelling develops fully in 24 to 72 hours:

The morphologic changes in the swelling are given below:

1. Accumulation of mononuclear cells around small veins or venules called perivascular cuffing and deposition of fibrin in the interstitium.
2. Oedema due to increased vascular permeability.
3. Presence of CD4 (helper) T lymphocytes in the swollen tissues as revealed by immuno peroxidase staining.
4. Replacement of perivascular infiltrated lymphocytes by macrophages within 2 to 3 weeks.
5. Transformation of accumulated macrophages in the swelling into epithelium like cells called epithelioid cells.

A microscopic accumulation of epithelioid cells surrounded by a color of lymphocytes is called **granuloma**.

A granulamatous inflammation is a characteristic feature of type IV delayed hypersensitivity. Transplant rejection and contact dermatitis are examples of type IV hypersensitivity. T cell mediated cytotoxicity is seen in this type of hypersensitivity. Cytotoxic T lypmphocytes (CTLs) which are also called CD8+T cells destroy antigen bearing target cells.

Autoimmune diseases

There are a few examples of autoimmune diseases in animals. The important features of these diseases are given in Table-12

Table 12: Immune diseases, their main pathologic changes and other features in the body.

Diseases	Animals affected	Pathologic changes and important features
(i) Allergic encephalomyelitis (Post vaccination encephalomyelitis)	Dogs	(i) Demyelinating encephalitis --a pathologic effect of post rabies vaccination in the form of tissue culture vaccine of tissue of the central nervous system from the same or different species. (ii) Infiltration of mononuclear cells in the brain and destruction of white matter and vasculitis noticed in the vaccinated canines.

(ii) Orchitis	Dogs	(i) An inflammatory disease in dogs due to immunization with spermatozoa or their antigens and released spermatozoa from injured testicles and some other infections. (i) Focal or diffuse infiltration of epididymis, vas deferens and testes with lymphocyts. Cellular infiltration of monocyts may also be seen in the testicles.
(iii) Pemphi vulgaris(an autoimmune disease caused by released proteases following interaction of auto antibodies with some intercellular substance in the skin and mucus membrane .	Dogs, cats and man	(i) Large ulcers and bullae of the skin. (ii) Epidermal cells separated from each other, loss of inter cellular bridges and degenerated inter cellular substance of epithelial cells. Disruption of incellular junctions epidermis by antibodies against desmosomes causes skin ulcers.
(i) Autoimmune hemolytic anaemia (AIHA)	Cattle horses, dogs, cats and mice etc.	(1) Hemolytic aneamia and red cell debris. The debris is removed by R.E cells of the liver and spleen. (i) Hepatomegaly and thumbocytopenia in the affected dogs. Intravascular and extravascular hemolysis produced. (ii) IgG and or IgM coated red cells or agglutinated red cells removed by microphases of the liver and spleen. (iii) Antierithrocytic antibodies bound to red cells activate the complement system leading to formation of membrane attack complex (C5-9) and intravascular haemolysis produced by C5-9 in patients like dogs.

		(i) Red cells are coated by IgG are phagocytosed microphases or mononuclear cells. (i) Marked by binding of the complement to the antibody-GBM complexes and activation of the complement
(v) Antiglobomerular basement membrane nephritis (anti GBM nephritis) A rare disease in horse due to autoantibodies against non-collagen domain of the type IVcollagen in the glomerular basement membrane.	Horses	(ii) Glomerulonephritis and pulmonary haemorrhage and tissue damage due to hydrolytic enzymes released from neutrophiles. (iii) C5A (a chemoattractant) arising from activation of the complement system causing infiltration of neutrophiles in the renal lesions. (iv) Presence of IgG and complement in the GBM and linear pattern formed by antibodies (Hb) in the globemerular basement membrane (GBM)
(vi) Arthus reaction (Arthus phenomenon – an example of type III hypersensitivity or immune complex disease) Preliminarily sensitized animals show Arthus reaction to repeated injections of a non-toxic antigen e.g horse serum. For example, the serum of a rabbit sensitized with horse serum is	Animals e.g. rabbits	(i) Marked by localized tissue necrosis arising from acute immune complex vasculitis. Antibodies in the immune animals react with specific antigens administer intracutaneously to produce local necrotic or inflammatory oedematous lesions. Antigens diffuse into vascular walls to produce immune complexes which are deposired locally.

administered into a rabbit which, thus, receives immune bodies passively. If the latter is injected 1-2 ml of the same horse serum, then it develops local hyper sensitivity reaction.	.	(i) Local lesions in the form of oedema, haemorrhage and ulceration in vascular walls and fibrinoid necrosis in the vessels. (ii) Complement, immunglobulins and fibrinogen stained within the venular walls by immuno fluorescent stains.

Chapter **8**

Oncology (From the Greek words: onkos-bulk, mass and logos-word)

Animal oncology deals with the study of tumours in animals. Tumours (neoplasms) are new growths beginning with transformation of normal tissue into neoplastic tissue which is characterised by atypical structure or function. Cellular proliferation in all non-neoplastic conditions (e.g. repair, regeneration and hyperplasia etc.) is arrested after the stoppage or cessation of the causative stimuli and the newly formed tissues differentiate and assume the properties of the tissues of origin and also maintain their connection with the surrounding tissues. Growths of these kinds are altruistic (i.e. for the well being of an individual) and are under the regulatory influences of the body of an organism. But tumours proliferate continuously and maintain progressive uncontrolled or un-regulated development with even parasitic behaviour.

Tumours have been defined variously by different pathologists but the definition givern by Willis (1960) seems to be quite satisfactory.It is given as follows :

"A tumour is an abnormal mass of tissue, the growth of which exceeds and is un-coordinated with that of normal tissue and usually persists in the same excessive manner even after the cessation of the stimuli which evoked the change."

Occurrence of tumours has been reported in animals, fishes, amphibians and sometimes in reptiles. Epithelial and connective tissue tumours are found in birds or chickens. Spontaneous tumours which arise under natural conditions are reported in almost all species of animals and mammals. Mammary cancer is very common in mice due to their increased susceptibility. Tumours rarely occur in dogs. Investigation on the incidence of tumours in animals is very much fascilitated due to improved methods of diagnosis. Chances of finding out tumours in humans or animals are quite encouraging with increase in life expectation (longevity).

Features of tumours

There are several features of tumours which quite distinguish them from normal tissues. These are as under :

1. Tissue atypicalness or anaplasia. Atypicalness which is noticed in morphological and functional properties of tumours arises from decreased differentiation of the tissue. Tumour cells which do not conform to or resemble any histological prototype in structure, arrangement or staining qualities are called anaplastic cells. This is an important feature of malignant tumours (e.g. cancer and sarcoma etc.)
2. Resemblance to healthy cells of the tissue of origin as usually seen in benign types of tumours.
3. Lack of orderly structural arrangement.

   Neoplastic cells cannot differentiate and organize in normal manner.
4. No useful function in the body.
5. Independent existence in the body. Dependence on the body is just for food or nutrition. When the patient dies, there is also an end of the tumour. It thus, behaves as a typical parasite.
6. Autonomous new formation of tissue. Autonomy is the most important feature of tumour. In short, it grows for its own sake following its own rules with no care to the host's well-being.
7. Continuous proliferation without control
8. Usually recognizable due to its grossly detectable swollen character. Tumour means a swelling in the body of an organism. In short, it occurs as a swelling or lump of tissues.

MORPHOLOGICAL ANAPLASIA

A variety of sizes and forms of the cells are found in a morphologically anaplastic tumour. Such cells have a lack of correspondence between their protoplasm and nucleus. The nucleus is large and has altered structure and reduced number of mitochondria. The natural arrangement of cells is disturbed. In anaplastic tumours of glands, lobes are sometimes, absent. They lack usual structure and the excretory ducts are infrequently absent and acini are abnormally formed in the glandular tumours to give rise to cystic structures.

CHEMICAL ANAPLASIA

The tumours are deficient in ash contents and have more water than usual, and have somewhat increased amount of potassium and sodium. Tumours are also deficient in calcium and magnesium. Fatty infiltration and increase in unsaturated fatty acids are observed in tumours. The amount of lipids, especially, cholesterol, is increased. There is an accumulation of lactic acid in tumours. The content of cystine, methionine and tyrosine is diminished in the tumours. The activity of enzymes like proteinases, aminopeptidases, transaminases, deaminases and phosphatases etc. is depressed. The activity of riboflavin, cytochromes and dehydrogenases is decreased in the tumours. While normal tissues die due to oxygen deficiency, tumours can exist due to anaerobic glycolysis. Lactic acid in tumours causes slight acidity of the environment leading to swelling of connective tissue fibres, growth and proliferation of the tumour cells.

There are one functional and two morphological features of malignant tumours. These are :-

1. Metastasis
2. Anaplasia
3. Invasion

METASTASIS

Metastasis refers to secondary growth of tunour somewhere in the body at a distance from the primary focus. Thus, tumours become multi-centric i.e. there is a simultaneous growth at several places in the body. The tumour cells at the primary and secondary sites resemble each other morphologicaly and functionally. Keratin whorls or acini in the tumour of glandular organ and colloid formation etc. are seen in the secondary tumours. Some of the properties of the original parent tissues are retained by secondary growths. The common sites of metastasis are regional lymph nodes, lungs and liver. Contact between two lips can lead to transfer of cancer cells from one lip to another.

ANAPLASIA

A histologic attempt to depict malignancy is an anaplasia. Anaplasia is characterized by following facts :

1. Lack of resemblance to the tissue of origin i.e. atypical appearance.
2. Disturbance in orientation of the cells in relation to one another.
3. Disturbance in polarity of the cells (i.e. improper orientation) of the structures or cellular components within the cells.
4. Deficiency in the differentiation and specialization of cells.
5. Cells of various shapes and sizes.
6. Presence of large number of mitoses as distorted spindles, tripolar mitoses and other nuclear abnormalities.
7. Deficiency of the connective tissue stroma.

INVASION (METASTASIS)

Malignant cells of the tumour grow in all directions from its site of original or primary growth along the lines of the least resistance or by penetrating into the blood vessels or lymphatics. Maligannt epithelial cells (cancer cells) show lymphatic permeation i.e. these cells enter through the walls of lymph channels. Maligannt connective tissue cells (sarcoma) penetrate through the walls of arterioles and venules and even larger vessels. Cancer cells entering into the lymph vessels or thoracic duct are ultimately carried to the neighbouring lymph nodes or lungs in which these cells grow to form secondary growths of neoplasms (i.e. metastatic growths). Malignant cells in the veins reach the lungs and get lodged there where as those entering in the arteries are arrested in the organs like kidneys. Later, secondary tumours are found in the lungs and kidneys etc. cancer cells grow and cover the mesothelial lining of the serous cavities as a continuous sheet. This phenomenon is called trans-coelomic dissemination. The extracellular matrix (ECM) includes basement membranes (BMs) and interstitial connective tissue. The ECM is made up of collagens, proteoglycans and glycoproteins. The tumour cells interact with the ECM. The cancer cells breach the basement membrane, traverse through the interstitial connective tissue for ultimate penetration of the vascular basement membrane to enter the blood stream which carries the cancer cells i.e. tumour emboli to grow in different sites in the body. These secondary growth of tumours at different sites are called metastases. In short, phenotypic attributes of malignant tumours include excessive cellular proliferation, local inasiveness and formation of distant metastases.

STRUCTURE OF TUMOURS

Tumours are organoid in structure i.e. they have parenchyma and stroma like normal tissues. The basic properties of tumour are associated with parenchyma. The parenchymal cells, e.g. hepatic cells in the liver, are called type cells of the tumour (e.g. hepatic adenoma) and the stroma of the tumour is a supportive structure of fibrous connective tissue elements containing the blood vessels to nourish the tumours.

Stroma

It is non-neoplastic part of the tumour and originates from normal tissues of the host. Proliferation of tumour cells stimulates connective tissue to multiply and the newly formed tissue has new vessels and collagenous material around or between the clumps of tumour cells. Stroma also grows from the pre existing connective tissue elements along with growing tumours in order to maintain the blood supply. The stroma is a vascular connective tissue in nature and will be in abundance but it is scanty in some tumours. Presence of collagenous tissue will be little with in a few blood vessels. It may contain elastic fibres, mast cells or may even develop islands of bone or cartilage (presumably due to cell metaplasia). Lymphocytes and inflammatory cells may accumulate in tumours due to inflammation and show cellular or humoral response (immunity against neoplasms). The blood vessels of the stroma remove the waste products of tumours. Malignant connective tissue products of tumour may produce extracellular material which may be excessive or slight. In fibroma, each tumour cell is supported from the next by the bands of collagen produced by the tumours.

Tumour cells originate from the cells capable of division. Malignant tumours can arise from fully differentiated cells, and cells of tumours can also differentiate as seen in cornifying squamous cell carcinoma. Fall in differentiation of tumour cells is directly related to the malignancy. When a cell gets greater degree of differentiation, it losses its tendency to divide with decreased telomerase activity.

Gross appearance

Tumours form enlargements or lumps which may reach very large sizes. Malignant tumours may not reach a conspicuous size. Infiltrative and metastatic growths (secondary tumours) are shown by malignant tumours. Some tumours do not form detectable primary swellings during autopsy and such growths are called occult neoplasms. Tumours form bulges in the organs or the parts involved in the body. Necrosis, haemorrhage and ulceration can be seen in the neoplastic growths. The tumours differ in colour and consistency from tissues surrounding such growths. These tumours may be white, yellow or black in colour. Cystic structures can be found in the tumours.

Microscopic appearance

The cells of the tumour resemble the cells from which they arise. (i.e. the type cells). The type cells are almost always of a single kind (for example, epithelium in a papilloma). Mixed tumours in the mammary glands show more than one kind of neoplastic cells. Anaplasia of a tumour refers to its quality of not conforming to any histologic prototype in structure and is seen in the malignant tumours. Tumours lack orderly structural arrangement as noticed in a fibroblastic tumour. Fibroblasts run in harum scarum (i.e. in a reckless way) in all directions on the cut surfaces of neoplasms. Infiltrative, invasive and metastatic properties are shown by malignant tumours and the only suitable basis of diffenentiation of benign tumours from malignant tumours is microscopic examination of the tissue sections of tumours. High degree of the cellularty of the tumour cells, nuclear hyperchromatism and numerous mitotic figures are the histological (i.e. litmus test) indicators of the malignancy.

NOMENCLATURE

Tumours may arise from epithelial, connective, nervous and muscle tissue etc. The names of the tumours are usually derived from those of the tissue from which the tumours arise and the suffix 'Oma' (e.g. fibroma, lipoma and neuroma). The suffix 'Oma' means tumour. The suffix "sarcoma" is used to indicate malignancy for all tumours of supporting tissue and the suffix' carcinoma' is similarly used for designating malignant tumours of epithelial tissue origin.The suffix "blastoma" is used for tumours originating from a stage seen in embryonal development of tissue or organ. In nomenclature, only Greek terms for cartilage i.e. chondros is used and by adding i.e. the suffix 'Oma' to the root of the word, the tumour called chondroma is constituted.

Classification

There are many ways or principles followed for classification of the tumours. Consideration of the behaviour or clinical factors along with the histogenesis of the tumour provides a very reasonable basis for classifying them. In short, the clinical features and histogenesis are two important bases (considered to be quite sound ones) for neoplastic classification.

The types of neoplastic classification are as under :

1. Histogenetic classification
2. Behaviouristic classification
3. Descriptive classification
4. Regional classification
5. Aetiological classification
6. Embryological classification

1. Histogenetic classification

Tumours can be classified on the basis of tissue of origin of the tumours (i.e. the type of the tissue from which a tumour originates). Histogenesis of the tumour seems to be very satisfactory basis for its classification. This system helps to form several groups or classes of tumour which are as follows :

(i) Tumours of epithelium.
(ii) Tumours of connective tissue.
(iii) Tumours of haematopoietic tissue.
(iv) Tumours of nervous tissue.
(v) Mixed tumours (i.e. tumours containing more than one kind of tissues).
(vi) Miscellaneous groups of tumours (i.e. tumours not fitting clearly into of the aforesaid groups).

Tumours grouped according to histogenesis resemble each other grossly and microscopically in a particular group whereas tumours of different groups tend to look different and also tend to behave differently. For example, a fibroma is grossly and microscpically different from a papilloma of epithelial tissue origin. Connective tissue tumours tend to look alike and behave alike but there are tumours whose tissue of origin (i.e. parent tissue) remains unsettled due to difficulty in determing the parent tissue. For example, Ewing's sarcoma is connective tissue tumour according to some pathologists whereas others consider it to be a tumour of haematopoietic tissue.

2. Behaviouristic classification

In this, the behaviour of the neoplasm is the main basis of classification which divides tumours into two groups, namely, (i) Benign tumours and (ii) Malignant tumours.

Benign tumours are localized and are much less dangerous. Malignant tumours severely injure the patient and can matastasise or cause invasion in different parts of the body of patients who ultimately die due to such neoplasms. The main distinguishing features of benign- and malignant tumours have been given in the following table (modified after Willis, 1960).

Table 12. Feautres of benign-and malignant tumours.

Features	Benign type	Malignant type
1. Structure	Often typical and resemblance to the parent tissue i.e. tissue of origin.	Often atypical. Anaplastic and differentiation of the tumour tissue is imperfect.
2. Mode of growth	Tumour growth is expansive in nature with formation of a capsule around it.	Tumour is both infiltrative and expansive. As a result, there is an absence of encapsulation.
3. Rate of growth	Mitotic figures in cells are scanty and growth of tumour is usually slow.	Presence of rapid growth and numerous mitotic figures in cells i.e. tripolar,quadripolar or multipolar spindles and tumour giant cells having polymorphic large nuclei or two or more nuclei.
4. End of growth	Tumours may disappear, retrogress or cease to grow.	Rare cessation of neoplastic growth which may progress up to death of the patient.

5. Metastasis	No Metastasis	Frequently seen in different organs from the primary neoplastic foci and infiltrative growths are very dangerous.
6. Clinical results	Position or growth of tumour in some vital organs of the body causes complications like obstruction, ulceration, infection etc. and excessive production of hormones may have dangerous or fatal effects on the host, for example, an excess of parathromone level in parathyroid adenoma causes hypercalcaemia.	

A neoplasm grows by an increase in the mass of tumour cells. When a tumour grows fast, the differentiation of the type cells is less perfect with the loss of normal morphological pattern. The tumours have unlimited power of growth and continue to grow. Tumors can grow in a relative isolation surrounded by a capsule. The tumours showing expansive growth are benign and less atypical in structure. Tumours with expansive growth move aside the structures confronting them and such compressed organs may suffer from dysfunction. Infiltrative and invasive malignant tumours invade the surrounding tissues and produce metastasis (i.e. growth of tumour cells at distant sites due to their transport through circulation from its primary site). Metastasis of the tumour cells in other organs happens through blood vessels and lymphatics and also depends upon the direction of the blood flow in the vessels. For example, malignant tumours in the stomach show metastasis in the liver. Metabolic disturbances and emaciation of the organisms (cachexia) are caused by malignant tumours. Malignant tumours are accompanied by haemorrhages and usually cause death of the organism.

3. Descriptive classification

Different kinds of tumours are noticed in the same region. It is possible to distinguish between these different kinds of tumour with the help of descriptive terms, otherwise, they would have the same name. For example, the basal cell carcinoma and squamous cell carcinoma arise from the epidermis, but the use of the term like basal, or squamous has helped to divide them into two kinds. These tumours behave differently in their growth and metastasize to the regional lymph nodes. Basal cell caronioma is known to grow slowly. The names of the scientists are also used to designate or divide the tumours into groups e.g. carcinoma of the kidney is called Grawitz tumour (Paul Grawitz 1850-1932) and another such example is Wilm's tumour arising from embryonic kidney.

4. Regional classification

Since different kinds of tumour arise in the same part of a body, a particular region of the body does not provide a satisfactory basis for classification. Knowledge about the region or the location of tumour is an important fact for a clinician but a classification like tumours of the stomach does not mean adequately for him. Tumours of the stomach may be benign connective tissue tumour or malignant epithelial tumour but lack of knowledge of their histogenesis and behaviour leaves a pathologist in darkness.

Aetiological classification

There is very little knowledge about aetiology of tumours Aetiology (i.e. causes of tumour formation) cannot offer a sound basis for its classification. If this principle is adopted, like tumours can be placed in different groups and unlike tumours would be grouped together. Different aetiological agents can cause the same kind of neoplasms and the same agent produces different kinds of tumours. For example, methylcholanthrene produces cutaneous haemangioma and carcinoma in the ducks and chickness respectively. Different chemical carinogenes produce carcinomas when these are applied externally to the epidermis of the skin. Thus, aetiology adds more to confusion instead of a solution in the classification of neoplasms.

6. Embryological classification

This system of classification is based on embryology. Many similarities between the malignant and embryonic tissues tend to attract one towards adoption of such system of classification but at the same time, these tumours differ greatly from the embryonic tissues.

Embryological classification can place similar tumours in different groups or even unlike tumours can be grouped together For example, squamous cell carcinoma of the mucosa of the lip is ectodermal and that of the oesophagus is entodermal. Such cancer of the cervix is mesodermal in origin. If embryological principle of tumour classification is accepted, a smiliar or same tumour in view of the tissue of origin (e.g. cancer in this case) can be categorized into different groups. That, this system does not offer a sound base of classification.

In short, histogenesis and behaviouristic pattern of the neoplasm seem to produce a very reasonable basis of classification.

Different kinds of tumours have been classified into two groups and given in the table-13 on the basis of their histogenesis and behaviour of character.

Table 13. Classification of tumours

Tissue of origin	Behaviour	
	Benign	Malignant
Histogenetic classification A. Tumours of one parenchymal cell types Group 'A" Epithelial tissue		
I. Epithelium		
a. Surface epithelium	Papilloma	Carcinoma
b. Glandular epithelium	Adenoma	Adenocarcinoma
c. Transitional epithelium	Papilloma	Transitional cell carcinoma
Group 'B' Non-epithelial tissue		
II. Connective tissue		
a. Fibrous tissue	Fibroma	Fibrosarcoma
b. Cartilage	Chondroma	Chondrosarcoma

c. Adipose tissue (Fat)	Lipmoma	Liposarcoma
d. Bone	Osteoma	Osteosarcoma
e. Mast cells	Mast cell tumour	Malignant mast cell tumour
III. Vascular tissue		
a. Endothelium of the blood vessels	Haemangioma	Haemanglosarcoma
b. Endothelium of the lymph vessels	Lymphangioma	Lymphangiosarcoma
IV. Muscle tissue		
a. Striated or striped	Rhabdomyoma	Rhabdomyosarcoma
b. Non- striated or smooth muscle	Leiomyoma	Leiomyosarcoma
V. Haematopoietic tissue		
a. Lymphoreticular tissue	None	Lymph sarcoma (Malignant lymphoma), lymphatic leukaemia
b. Myelic tissue Myeloblasts and Erythroblasts	None None	Myeloid leukaem ia and erythroid leuk-aemia
VI. Pigment forming tissue		
Melanocytes	?	Melanocyte
VI. Nervous tissue		
a. Glial tissue	Glioma	-
b. Menings	Meningioma	
c. Neuron	-	Neuroblastoma
d. Nerve sheaths (Schwann cells)	Neurilemmoma (neuro- fibroma, Schwannoma)	Neuroblastoma Neurogenic sarcoma (Neurofibro-Schsarcoma or Schwannoma)
B. Tumours of more than one neoplastic cell type i.e. tumour cell types derived from more than one germ layer e.g. teratoma arising from totipotential cells.		
VIII. Several tissue components		
Multipotent cells		
a. Ovary	Teratoma	Malignant teratoma
b. Mammary gland	(Dermoid cyst)	Mixed teratoma)
c. Kidneys	-	Embryonal nephroma (Wilm's tumour)
d. Salivary glands	-	Mixed tumour of salivary glands
e. Sweat glands	-	Mixed tumour of sweat glands

Aetiology of tumours

There are different theories concerning the aetiology of tumours. But no single common cause for all kinds of tumour is known as of today. Moreover, many more causes of neoplasms are coming into light as the knowledge on this aspect is extending from day to day.

The important causes of neoplasms are as follows :

(i) Predisposing causes

(ii) Direct causes

(i) Predisposing causes (Indirect causes)

Family, strain, breeds, individual factors, species, age, sex and colour etc. are known to enhance the susceptibility of an organism to neoplasia. Examples of neoplasms are quoted below against each of the aforesaid factors :

Indirect cause	Tumours
Family or strains	Spontaneous tumours are frequent in strains of mice. Mammary adenocarcinoma in certain suspectible strain of mice is quite high.
Breeds	Tumours affecting Danish mice cannot be transplanted into English mice. High incidence of ocular carcinoma is noticed in the Hereford breed of cattle.
Susceptibility or Heredity	High incidence of tumour in susceptible individual. Heredity has some importance in neoplastic growth. Susceptibility is transmitted from one generation to the next. The susceptibility of an organism is an important factor in tumourigenesis. It is transmitted from one generation to the next generation. (e.g. certain inbreed mice). Some strains of mice show high susceptibility to mammary adenocarcinoma.
Species	Rabbits are less susceptible to neoplasia than mice. Incidence of some tumours can be linked to a particular age of an organism.
Age	Cancer is a disease of old age and is seen in the human age group of 65 in Western countries.
Colour and sex	Melanoma is common in grey horses. Sex also governs the incidence of tumours. Breast cancer or cancer of cervix is seen in women but seminoma is seen in males.

There are also other factors governing occurrence of tumours . Low calorie food retards growth of tumours. Endocrine glands affect neoplastic growth. The somatic hormone of the hypophysis stimulates growth of tumour. Estrone accelerates the growth of neoplasms of the mammary gland and genitalia where as testosterone and progesterone (a hormone of the corpus luteum) inhibits such neoplastic growths.

2. Direct causes of neoplasia

There are several factors which can convert normal tissue into neoplastic tissue. Some of the causes of neoplasis are as under :

I. Foetal rests (Foetal residues or Cohnheim's sembryonic theory)

Tumours can arise from foetal residues or foetal inclusions.

Sometimes, tumours originate from islands of unmatured or undifferentiated tissues, or immature tissues which have been misplaced or entrapped in other organs (not normally containing such tissues) Such misplaced cells start growing during postnatal life to develop into neoplasms and these tumours do not resemble the parent tissues and are said to be heterotopic tumours. A part of adrenal gland within the kidney can grow into a tumour called hypernephroma. Teratomas can, thus, grow from the misplaced or embroypnic tissues in different parts (e.g. ovary and testicle) of the body. These tumours contain different histological types of tissues. A cornified squamous cell carcinoma can develop in gall bladder.

II. Chronic Irritation (Theory of irritation)

Tumours can arise from a long continued action of various irritants on the tissues. Tumours may arise as a result of different traumatic injuries to tissues, chronic inflammation and other kinds of irritation to the tissues. Cancer has been noticed at the sites of old scars, ulcers and erosions and in the gall bladder in the presence of calculi therein. When burns heal very slowly, some highly prolific cells appear by natural selection and survival of the fittest at such sites. Such cells grow rapidly with uncontrolled power of proliferation to ultimately give rise to a neoplasm. Cancer in people using kangari in India or in the lips of pipe smokers appear from irritation which is suspected to cause such neoplastic growth. But it cannot be considered entirely satisfactory. This theory merely reveals the fact of a mere connection between a long acting irritant e.g. heat and growth of tumours. It is impossible to say whether or not the tumours arise from the protracted action of the irritant on the tissues. Chronic proliferative or regenerative processes are possible foci for appearance of tumours. Fibrosarcomas and malignant fibrous histiocytomas in the subcutis of neck and skeletal muscle of cats vaccinated with feline leukemia virus and rabies virus are examples of neoplasms induced by chronic inflammation.

III. Actinic light rays

Exposures to sunlight or actinic light rays can induce neoplasms on the unclothed parts of the body. Ocular carcinoma in cattle (Herefore breed) seems to originate due to action of the brilliant sun light. Ultraviolet light is a well known proven cause of cancer.

IV. Irradiation

Roentgen rays (X-rays) produce cancer in persons exposed to irritation. The exposed skin shows dermatitis and ulceration with ultimate development of cancer at such sites. Radiant energy e.g. ultraviolet rays and ionizing radiation (electromagnetic X-rays and α rays) and particulate radiations (α-particles, protons and neutrons etc.) cause malignant neoplasms like leukemias. Leukemias develop in humans exposed to radiations of atom bomb. But the cases of malignant neoplasms are rare in animals. Ionising radiation causes direct mutations of proto-oncogenes in the cellular genes (i.e. formation of c-oncs) The genes that regulate proto-oncogene expression also undergo structural alterations. Suppressor genes (anti-oncogenes) may also fail to check expression of cellular oncogenes

Chronic osteitis or osteosarcoma is produced in bone owing to localization of radioactive radium and thorium in their substance. Uranium is also known to cause cancer in the organism exposed to its action.

V. Chemical theory (Chemical carcinogens)

Carcinogens

Several chemical factors are known to produce neoplastic growth. The carcinogenic substance gains enterance into the body of an organism from without or have been possibly formed in the body as a result of some changes in the metabolism. Carcinogens act slowly. At first, there is no change. A long latent period follows it. Sometimes, tumours grow at places other than the sites of administration. For example, application of 1:2:5:6 dibenzanthracenne to the skin may cause growth of a tumour in the lung or mammary gland. Such growth in the lung indicates a general action of the carcinogenic substance. Pathogenesis of neoplasia due to chemical carcinogens can be considerd in different stages such as initating, dormant and promoting stages. In the initiating stage, normal cells are irreversely converted into latent or dormant tumour cells because of DNA damage or mutation. The next stage is promoting stage in which dormant cells pass on to the progressive stage of growth. Co-carcinogen (a non-carcinogenic substance like croton oil) on its application to the site previously treated with a chemical carcinogen, induces a rapid formation of neoplasms. An established tumour arises from promoting action. Use of anticarcinogens like mustard gas, inhibits the development of tumours i.e. promoting action is depressed.

Soot and gaseous products of combustion have carcinogenic substances causing chimney sweep's cancer. Papilloma and carcinoma can be produced in rabbit's ear after repeated application of the tar to the skin. Urethane produces cancer in the lungs of mice. Mode of administration of the carcinogens influences the kinds of neoplasms to develop in the experimental subject. O-aminoazotoluene (a dye) can be administered into animals in a variety of ways but the ultimate result is formation of only one kind of tumour (i.e. an epithelial tumour) in the body of the experimental animal. It indicates that the carcinogens have a predilection for a particular tissue or organ irrespective of the route or the site of administration.

The pathogenesis of all tumours cannot be explained on the basis of chemical (carcinogenic) theory (for example, the cause of so called occupational cancer).

The important chemical carcinogens are as follows :

1. 1:2:5:6: dibenzanthracene.
2. Methylcholanthrene.
3. 3:4-benzpyrene.
4. 9:10 – dimethyl – 1:2- benzanthracene.
5. O – aminoazotoluene.

To sum up, the important steps of chemical carcinogenesis (Fig. 30) are

1. Initiation
2. Promotion

The aforesaid concepts arose from experiments carried on mouse skin. Initiation is caused in the cells by their exposure to an appropriate dose of a chemical carcinogen. Initiation causes permanent DNA damage i.e. mutation in the exposed cells. This change is rapid, irreversible and has a memory. Promoters are such chemicals, which induce neoplastic features in the initiated cells. These promoters are nontumourigenic and do not affect the DNA. The changes induced in the cells by promoting action are reversible. Tumours fail to develop in the cells exposed to promoters. There are two types of carcinogenic factors, namely (i) direct acting carcinogens e.g. dimethyl sulphate β propiolactone and (ii) indirect acting factors (called procarcinogens) e.g. Benzanthracene and benzo (a) pyrene and dibenz (a,h) anthracenl and other substance like, aflatoxin β_1, fungicides and insecticides etc. Direct acting carcinogens act directly on the cells, bind to DNAs and convert the cells into cancer cells by causing permanent DNA lesions in the initiated cells. The initiated cells proliferate and undergo additional mutations to turn normal cells into malignant cells.

There is also a metabolic conversion of certain substances into carcinogens in the body and normal cells exposed to them are transformed into malignant cells. In chemical carcinogenesis, there is a mutation which affects oncogenes, cancer suppressor genes and genes regulating apoproteins. DNA is a primary target in this type of carcinogenesis. Ras mutations are seen in chemically induced tumours in rodents.

VI. Hormones

Estrogen produces adenocarcinoma in mice. It even causes cystic glandular hyperplasia in the mammary glands. Androgens are known to inhibit mammary carcinogenesis. Estrogenic hormones produce hyperplasia and neoplastic structures in the interstitial tissues of the testes of mice. Carcinogenic properties of many other hormones are not known.

VII. Biological factors

Some parasites and viruses are known to cause tumours in animals.

Parasitic theory (Parasites)

These may cause irritation at the site of their habitat where neoplasms develop. *Schistosoma haematobium* causes cystitis and cancer in the urinary bladder of human beings. *Cysticrecus fasciolaris* produces sarcoma in the liver of the rats which get infected by eating faeces of the rat containing eggs of *Taenia crassicollis.* Sarcomas develop around its cystic stages in the liver of rats.

Spirocerca lupi

It invades the wall of the stomach and oesphagus in dogs and spheroidal tumours are formed in their walls. These tumours are either fibrosarcomas or osteosarcomas. Bone in such tumour orginates from metaplasia of fibrous tissue into bone in the mass of tumour.

VIII. Viral theory

The important steps in normal cell proliferation include detachment of the cells from one another and binding of growth factor to a specific receptor on the cell membrane and some activation of the growth factor receptors. The signals emanating from the cell surface cause initiation and termination of cell divisions. Chemical messages arising from the cell membrane are transmitted to the nucleus across the cytosol. Nuclear regulatory factors or genes are activated for initiating DNA transcription and cell division. These regulatory genes control activation and suppression of transcription of proto-oncogenes There are primary and secondary messages arising from the cell membrane and cytosol respectively. These messages involve generation of enzymic reactions in molecules in the plasma membranes. The enzymic reactions cause phosphorylation of certain proteins which enter the nucleus to bind to specific portion of the nuclear DNA for initiation or termination of cell division. The genes causing normal cell division and differentiation are called proto-oncogenes. These cellular genes causing normal cellular proliferation and differentiation are precursors of viral oncogenes (v-oncs). That is why these are called proto- oncogenes. It is noteworthy to mention that retroviruses i.e. oncorna viruses transform the cells into malignat cells (cancer cells) during productive non-cytocidal infection with the feature of uncontrolled proliferation and the viral RNA is also transcribed from the integrated DNA or DNA intermediate.

The first event is a DNA damage in normal cells by injurious agents like chemicals, irradiation (x-rays) and viruses etc. DNA damage in vivo includes single strand or double strand breaks, loss of purine bases and chemical changes in the bases. Failure of genes affecting DNA repair, cell growth and apoptosis are conducive to mutations in the genome of somatic cells. If there is no successful DNA repair in the cells, DNA damage results in the mutation of DNA. These mutations in the DNA in normal cells cause permanent heritable features in the transformed newly proliferated cells. Inactivation or loss of cancer suppressor genes, (e.g. p-53 dependent genes not activated) after DNA damage, and alterations of genes regulating apoptosis and activation of growth promoting oncogenes lead to expression of altered gene products and loss of regulatory gene products. p-53 gene (i.e. guardian of cellular DNA) senses DNA damage and assists in DNA repair by inducing repair genes. It also directs cells with damaged DNA to undergo apoptosis. Hypoxic tumor cells also undergo apoptosis. Hypoxia activates p-53 gene which prevents tumour growth. If there is a homozygous loss of p-53 gene, there is no repair of DNA damage in the cells and the mutations become fixed in proliferative cells which undergo malignant transformation. The mutated cells, then, show clonal expansion of neoplastic cells, additional mutations in cells and heterogeneity ending in the formation of malignant neoplasms. Most tumours show monoclonal expansion i.e. a clonal expansion from a single progenitor cell affected with a genetic damage.

To sum up, RNA viruses like sarcoma virus (RSV a cause of cancer in chickens), avian myeloblastosis virus and RNA viruses in rodents bring about permanent heritable transformation of normal cells in malignant forms (cancers). A permanent genetic change in otherwise normal cells produces a cancer cell. Such a change in cells is called malignant transformation of cells. Viruses are known to cause tumours in animals. Special globular viruses like particles visible by an electron microscope are found in the filtrate of some tumours and neoplasms develop in animals which have been given such viruses. These viral particles are capable to cause foci of cellular proliferation in the body. Viral theory can not be applicable to all causes of neoplasia and it is difficult to accept filterable visuses as the general causes of tumours. Cancers in animals are caused by several RNA and DNA viruses. Bovine papilloma virus causes both benign and malignant tumours. Papilloma virus of the family Papovaviridae (DNA viruses), avian leucosis viruses (a RNA virus), mouse tumor, mammary tumor virus, murine leukemia virus and bovine leukemia virus of the family Retroviridae cause leucosis or leukemia in birds and animals. In transforming retroviral infection of animals like rodents a double stranded viral DNA (called provirus or DNA copy) produced by an enzyme called reverse transcriptase is permanently inserted into the cellular DNA genome. in other words, the complementary copy of the viral DNA is integrated into cellular DNA and one sees the presence of viral genome in the host cell genome. Host cell transcriptase causes ranscription of complementary viral RNA i.e. positive (+) messenger RNA responsible for coding viral proteins. As a result, the normal cells are transformed into malignant cells with permanent heritable features.

Some of the important viral diseases characterized by neoplastic transformations are the following :

Diseases	Causes	Animals affected
Avian leucosis complex	A RNA virus	Chickens
Marek's diseases	A DNA virus	-do-
Rous sarcoma	A RNA virus (i.e. RSV)	-do-
Bovine papilloma	A DNA virus	Cattle
Oral papilloma	-Do-	Rabbit
Shope's papilloma	-Do-	Cotton tail rabbit
Tumour of the parotid gland (caused by polyoma virus)	-Do-	Mice (murine)
Warts	-Do-	Cattle, horses and dogs
Sarcoma	A RNA virus	Cat
Leukaemia	A DNA virus	Mouse

IX. Milk factor

Milk factor is a high molecular protein with a ribonucleic component and viral properties. High incidence of mammary carcinoma is found in certain strains of mice (up to 80 percent). If new born mice from the strain with high incidence are fed by females (foster mothers) of a resistant strain with low incidence of mammary cancer, incidence of the mammary cancer falls down in the new born mice. This fact of high incidence of mammary cancer is linked to or dependent on the presence of some substance in the milk known as milk factor in the maternal milk of the strain of highly susceptible mice.

X. Oncogens (Cancer causing genes)

Oncogenes refer to genes capable of producing tumourigenesis under certain conditions in an individual. Such genes are normally found in cells of human and animals and more than 30 of protooncogenes have been identified. The proto-oncogenes are precursors of oncogenes in the cells. The oncogenes have been named as mos (mos for the maloney murine sarcoma) or mye (for the tumour of birds known as myelocytoma). An approporiate length of the deoxyribonueleic acid from a tumour is incorporated into the genome (having genetic information) of the cells which get transformed or rendered tumorogenous. Such genes with ability to form cancer or tumour are called oncogenes. Oncogenes have been noticed in lymphoma, leukaemia, sarcoma, glioma and neuroblastoma in humans. The oncogenes are carried by retroviruses (oncornaviruses) and do not play any role in replication of retroviruses. Oncogenes are altered cellular genes but not viral genes and are copied by the virus from a mammalian gene.

It is not possible to propose either a single living virus or a chem.cal carcinogen for genesis of all neoplasms. Some specific genes (called oncogenes) have been assigned certain steps in tumorogenesis. In spite of some knowledge on chemical carcinogens, oncoviruses and oncogenes, the transformation of a normal cell into cancer cell is not clearly understood as yet and as such, one can not postulate a single pathway by which all forms of cancer can develop into an individual. Multiple genetic alternations are known to impart a malignant status to a normal cell. Irreversible neoplastic transformation involves changes in the genomes (i.e. the composition of the DNA of the nuclei). Gene mutation for the tumour induction has been noticed and many mutagenes active on Drosophila produce carrcinogenesis in mice. Mutation is a popular theory of carcinogenesis in mice. The chemical carninogens cause mutation by binding to deoxyribonucleic acid and also prevent its normal replication. Radiation is also known to cause mutation in the cells. Oncoviruses are known to bring about mutation by binding their deoxyribonucletc acid to that of the cells and the whole process results in alteration of genetic composition with resultant tumorigenesis. Oncogenes are known to drive cells to uncontrolled proliferation whereas tumours suppressor genes normally restrain cell proliferation. A neoplasm can not develop from a mere presence of an oncogene in a cell. Oncogene is activated and causes normal cells to divide in an uncontrollable manner. Oncogenes remain inactive or dormant in the cells and rather regulate normal cellular growth. Gab2 is a signaling protein molecule that underpins normal responses as well as oncogenesis in the cells. It is involved in transmitting signals from the cell surface to the interior of cells to undergo the process of division and migration etc. This molecule (Gab2) can be switched off in order to prevent transmition of proliferation signals to combat Gab2 activated conditions i.e. breast cancer and some kinds of leukemia. When there is no need of the role of Gab2 molecule, it usually remains switched off. In the cultured cells, a mutation changing one aminoacid causes transformation in the cells. i.e. tumorigenesis and a gene is transcribed rapidly by a chromosomol rearrangement. Translocation of an oncogene from chromosome 8 to chromosome 14 is noticed in Burkitt's lymphoma. Oncogenes act in concert with many other factors for tumour induction. Sudies on the tumourgenic retroviruses (i.e. oncogenic RNA viruses) in animals led to discovery of many oncogenes behaving as passengers in these viruses.

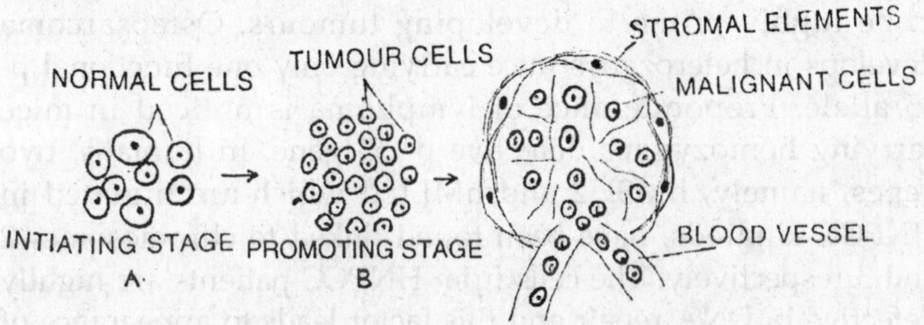

Fig. 30 Two-stage- mechanism of tumorigenesis

Gene amplification refers to an accumulation of a gene out of proportion to a cell's normal chromosome number. One can even consider it a mechanism by which cells can acquire and increase expression of genes. Gene amplification is increased in malignant cells. Oncogenes amplification, a genetic alteration, contributes to proliferative propensity in a clone of tumour operation. Genes capable to induce neoplasia are normally present in the cells in the repressed state. Due to failure of repression, the onocogenes become active to lead to the state of neoplasia. In two stage mechanism of tumourigenesis, initiating factors cause mutation and the promoting agents activate the altered genes to neoplastic transformation of cells (Fig.28). A recently identified killer gene has been found to make a protein which is an essential part of enzyme called telomerase with propensity to drive cells to grow without restraint to induce carcinogenesis. Tumour suppressor genes are known to suppress the ability of the cancer cells to form tumours. Somatic mutation inactivates tumour suppressor genes either by changes at the DNA sequencing level or by gross rearrangement. p- 53 and Rb genes are examples of tumour suppressor genes. The hereditary retinoblastoma gene (Rb) is considered a prototypic tumour suppressor which is a nuclear phosphoprotein with central role in cell cycle regulation. p-53 restrains cell growth and any mutation in this gene can lead to unregulated cell growth (neoplasia). It is also a key regulator, which also controls expression of other genes. One can both insert exogenous genes and delete normal genes from mice by transgenic technique. Mice lacking a functional p- 53 are found to be highly prone to developing tumours. Osteosarcoma develops in heterozgoys mice carrying only one functional p-53 allele. Preponderance of lymphoma is noticed in mice carrying homozygous defective p- 53 gene. In humans, two genes, namely, hMSH2 and hMLIHI which are mutated in HNPCC kindreds, have been found linked to chromosomes 2 and 3 respectively. The cells from HNPCC patients are hightly defective in DNA repair and this factor leads to appearance of cancer is such families. Role of p53 mutation in evolution of some soft tumour tissue has been recognized and the abnormal genes detected per total tumours are as follows :-

Tumour	Abnormal genes per total Tumours
Ewing's Sarcoma	2 of 8
Fibrosarcoma	1 of 2
Leomyosarcoma	2 of 6
Liprosarcoma	1 of 4
Rhbdomyosarcoma	2 of 6

Specefic cytotoxic T cells against antigens i.e. peptides derived from mutated p53 have been noticed.

Rb gene mutations (Retinoblastoma gene mutations)

Total partial gene deletions appear important mechanisms of Rb intractivation in certain sarcomas while during studies of the Rb gene at the gross structural level in DNA and RNA in soft tissue sarcoma, Rb gene deletions were noticed in two of 5 malignant fibrous histiocytomas and 2 of 4 liposarcomas. In Ewing's sarcoma, a solid tumour chromosome transformation has been noted and there is a chromosomal recurrent (11,22) in both osseous and extraosseous tissues of Ewing's sarcoma.

Oncogens are the genes of malignant tumours (cancers) derived from protooncogenes which considered to be first formed cellular genes. Protooncogenes (i.e cellular genes) show important roles in normal cell growth , multiplication and differentiation of newly formed cells. A mutated protooncogene marked by alterations in its structure causes neoplastic transformation of cells. This altered or mutated protooncogene is called a cellular c- oncogene (c- onc) In others words, there is a proviral insertion (DNA copy or DNA step) near a protooncegene in the cellular DNA of the infected cells with cancer causing RNA viruses called oncornaviruses. Proviral insertion causes a structural change in the cellular gene. As a result, this cellular gene is converted into a cellular oncogene (c-onc). Permanent heritable features of malignancy are noticed in the proliferated or transformed cancer cells. Oncoviruses e.g. oncornaviruses of the family Retroviridale are tumourigenic viruses where as oncogenes are mutated cellular genes i.e. mutated protooncogenes with capability to form tumours in organisms.

Viral oncogenes (v- oncs) are peculiar transforming RNA sequences in certain viruses e.g. retroviruses causing neoplasms in animals. V-onc sequences in the genomes of transforming retroviruses are almost identical to sequences noticed in the DNA of the normal cells. The sequences of cellular genes which are noticed in genome of transforming retroviruses are designated as viral oncogenes (v-oncs). Since retroviral oncogenes (v-oncs) were primarily detected as viral genes, proto-oncogenes are named after these viral homologs. These cellular genes which are noticed in the genome of transforming retro viruses are designated as viral oncogenes. Such observations indicated the process of transduction of retroviral genes (v-oncs) through a recombination (that is an exchange of genes between DNAs of chromosomes of host cells with oncorna-viruses). A v-onc is named after the type of the malignant virus in species of animals. For example, v-oncs present in feline sarcoma virus and simian sarcoma virus are designated as v-fes and v-sis respectively. When the prefix 'V' is dropped, the terms fes and sis indicate the corresponding protooncogenes in the cells. Leukaemias in rodents are examples of malignant transformation of cells by cancer causing RNA viruses . Oncogenic DNA sequences or viral oncogenes (v-oncs) are also found in oncogenic DNA viruses causing cases of spontaneous cancers. Proteins encoded by oncogenes in the cells are called oncoproteins which bear some resemblance to products of proto-oncogenes.

To sum up, protooncogenes become oncogenic by retroviral transduction (v-oncs) and the transforming genes of the retroviruses capable to induce neoplasms in the body are belived to have evolved through capture (transduction) of sequences from cellular genes called protooncogenes (c-oncs). Even certain influences alter the behaviour of protooncogenes and convert them into cellular oncogenes i.e. c-oncs.

The important features of oncoproteins encoded by oncogenes are:

1. Lack of regulatory proteins in oncoproteins.
2. Production of oncoproteins in transformed cancer cells (independent of growth factors).

Protooncogenes encode proteins for promoting cell growth in normal cells and also mutate to form cellular oncoogenes (c-oncs.)

Some examples of proto-oncogenes are as follows:

1. N-myc and myc (human tumour expressing gene products).
2. fos (v-oncs present in murine ostosarcoma virus).

3. fes (v-onc in feline sarcoma virus).

The normal regulatory genes are :

1. Growth promoting protooncogenes
2. Growth inhibiting cancer suppressor genes (anti-carcinogens)
3. Regulatory genes of apoptosis or apoptosis genes.

These three classes of genes are the main targets of genetic damage. The DNA repair genes influence the reparative processes involving the repair of damage in genes like proto-oncogenens, tumour suppressor genes and regulatory genes of apoptosis. Failure of DNA repair may be followed by mutation in the genome i.e. DNA. Thus, DNA repair genes act like tumour suppressor genes. Growth factors stimulate normal cells to proliferate. But there are certain growth factors which participate in tumourigenesis. Genes that encode growth factors undergo mutations to become oncogenes. For example, proto-oncogene- c-sis which encodes the growth factors PDGF-α chain (platelate derived growth factor â chain) undergoes mutations to be oncogenic and shows overexpression to form a tumour called astrocytoma.

A gene called Myc in its mutated form produces a Myc protein which gives signal to cells to divide uncontrollably in genetically modified mice. If the mice are fed doxacycline, the mutated Myc gene gets turned off. As result, there is a stoppage of protein flow and cellular proliferation. This explains how cancer can be switched off in the body of organism.

The steps in transformation of proto oncogenes into oncogenes are as follows:

1. Structural changes in the genes

These changes lead to synthesis of oncoproteins (abnormal gene products having abnormal functions).

2. Changes or disturbances in regulation of gene expression.

These changes increase production of growth promoting proteins.Thus, protooncogenes are affected by structural and regulatory changes. Together ras and myc transform fibroblasts into malignant forms.Multiple genetic alterations arise from activation of many oncogenes and loss of cancer suppressor genes or anti -oncogenes A ras point mutation is noticed in human pancreatic adenocarcinoma and cholangiocarcinoma. Translocations and inversions are processes which activate proto-oncogenes by chromosomal rearrangements. Burkitt lymphoma is marked by translocation induced overexpression of proto-oncogenes. Translocations are caused by the movement of the c-myc containing segment of chromosome 8 to chromosome14q band 32 in Burkitt lymphoma.

Chromosomal rearrangements in cells activate proto-oncogenes. Specific translocations cause an overexpression of proto-oncogenes in lymphoid tumours. Activation of proto-oncogenes and overexpression of their products arise from reduplication and gene amplification of the DNA sequences. Oncogenes and oncoproteins are considered as altered versions of their normal counterparts.

In neoplasia, information reaches the nuclei through a transduction pathway and influences responder genes that initiate the influenced cells to undergo mitosis (i.e. DNA replication and cell division). DNA replicates in S-phase of mitosis and interphase of the meiosis. Cells undergo malignant transformation when the genes encoding transcription factors mutate. There are several protooncogenes which act like transcription factors for regulating the cell proliferation. When transcription factors, say, protooncogenes mutate, malignant cells (cancer cells) are produced. Tumour suppressor genes include genes like p53 annd Rb gene (retinoblastoma gene) of which mutations are followed by malignant transformation of cells i.e. cancer cells. Oncoproteins i.e. products of myc, myb and mutant ras protein and fos oncogenes are localized in the nuclei and can cause uncontrolled proliferation of cells without any external signal or stimulant. Certain growth factors involved in proliferation of normal cells causing cell growth e.g. granulation tissues are also involved in some tumourigenesis. If the genes encoding growth factor undrergo mutation, they become oncogenic. Protooncogene cis produces â chain of growth factor called patelet derived growth factor (PDGF). Astrocytomas and osteosarcomas produce PDGF. Carcinomas produce the transforming growth factor (TGFá). The ras proteins are signal tranducing onco proteins in the nuclei. The mutated ras gene is a dominant oncogene in the colon and pancreatic cancer of humans. Certain tumours produce growh factors which stimulate them to proliferate and cause abnormal cellular masses. Signal transduction system converts the extracellular signal into intracellular signal. Ras gene in the cells is activated through growth factor receptors. The inactive GDP-bound ras recruits Raf -1 and stimulates and uses MPP kinase pathway to send growth promoting signals to the nuclei. The mutated ras in activated state causes continous proliferation of cells i.e. the state of neoplasia

Immune surveillance/Tumour immunity, Class I MHC (major histocompetibility complex) molecules, tumour specific antigens (TSA s) and (tumour associated antigens) TAAs

Tumour cells in the body are destroyed by an immune reaction. Activated T cells destroy tumour cells be secreting lymphotoxins or by binding to the tumour cells. Tumour cells are engulfed by activated macrophages. Neoplasms like cancer are seen in old people due to weakened immune system. Age factor also seems to play its role in carcinogenesis. Tolerance to antigens of the virus induced tumours differs from those of chemically induced tumours. Different specific antigens are present in the tumours developed at different sites in the skin of an animal following injection of a carcinogen like methylcholanthrene. Specific antibodies in some cases of tumours in the individuals have been described. However, theory of immune surveillance is not fully clear or established as yet. It means the recognization and destruction of non-self (foreign) tumour cells following their appearance in body. Occurrence of the tumour in an individual indicates that the immune surveillance is not perfect or fully developed in the host body. Many tumours in the body escape the policing of immune mechanism (i.e. surveillance) and this fact leads to death of patients. Tumour antigens have been detected in experimental cases of tumours and in some malignant tumors (cancers). Tumour specific antigens (TSAs) occur only on tumor cells but not on normal cells in the body. Tumour associated antigens (TAAs) occur in neoplastic cells as well as normal cells in the body. TSAs have been noticed in chemically induced tumours in rodents. T cell receptors of cytotoxic CD8+ T cells recognize peptides (antigens) with in tumour cells bound to major histocompatibility complex (MHC) molecules e.g. class 1 MHC molecules and participate in destruction of tumour cells.TSAs cause a cytotoxic response to destroy the neoplasms.

MHC class I molecules are synthesized within the cells and these molecules bind to antigenic peptides derived from proteins e.g. peptides derived from mutated forms of cellular proteins or mutated proto-oncogenes and suppresson genes tumour cells etc. Class I MHC molecules present these peptides originating within the tumour cells on the the neoplastic cell surface. Even viral peptides of the cells infected with oncogenic viruses are also presented on the cell surface by MHC class I molecules. Normal cells reveal TAAs on their surfaces. TSAs are derived from peptides within tumour cells and these are transported by class I molecules bound in their antigens binding clefts to cell surface. Ultimately these are presented to CD8+ T cells for their destruction. T cell receptors of these cells recognize these peptides (i.e. a process of antigen recognition) bound to the antigen binding clefts of MHC molecule and a cytotoxic effect is induced by CD8 + cytotoxic cells which secret some chemicals to kill the tumour cells. TSAs expressed on the surface of tumour cells also induce a humoral immune response. A cellular immune response induced by TSAs is powerful in destroying the tumour cells. These CD8 +T cells are noticed in or in the vicinity of tumours. These T cells act effectively to kill virally induced tumours. Interlukin- 2 (IL-2) activates a population of lymphocytes which destroy tumor cells. The cellular response by cytotoxic T lymphocytes and NK (natural killer) cells are examples of cellular immune response. Both cellular and humoral responses act in concert to kill tumour cells. The humoral response is called antibody dependent cellular cytotoxcity (ADCC). ADCC involves binding of the Fc receptors of cells like monocytes, neutrophiles or non-sensitised cells with the Fc fragments of IgG antibodies which form a coating around the target cells. Oxygen free radicals (reactive oxygen species) generated by macrophages and cytokines secreted by these macrophages also destroy neoplastic cells. Tumour immunity is not very strong in the body because of lack of specific foreign antigens in tumours, inadequate or weak immune response and soluable antigens –antibody complexes interfering with cellular immunity.

Certain parasites are known to produce tumors in man and animals. *Schistosoma haematobium* produces bladder cancer in humans where as *Spirocerca lupi*, alveolar hydatidosis and bladderworm *Cysticercus fasciolaris* produce malignant tumor in animals. Tumor immunity in parasitic diseases is not fully understood and parasites and their larval developmental stages do not seem to elicit tissue growths (tumours) bearing foreign TSAs (tumour specific antigens) and tumour associated antigens (TAAs). An effective immune system is not illustrated in infected animals with parasites causing tumourigenesis.

Summarising tumours elicit some reactions (i.e. immunogenic) in the bodies of individuals. This fact can be used to protect the animals or living organismas against the neoplasms. Tumour specific antigens exist on the neoplastic cells as uncovered antigens. The antigens in or on the tumour cells can be induced by chemical carcinogens and are found to be distinct from each other. These could be found within the same animals, or can be caused by the same chemicals. Tumour antigens are also induced by the virus and can show humoral or cellular responses.

Tumours can be infiltrated with lymphocytes (say, T cells) which can be found in or around the neoplasms. Body defences against neoplasms are weak due to the following facts.

(1) Lack of foreign antigens on tumours.

(2) Too weak immune system.

(3) Soluble blocking factors like antigen anti-body complexes causing in- efficiency of the defence or immune system.

A difference has been noticed in specificity between tumour antigens of virus produced tumours and those of chemically induced tumours. All the induced tumours by a particular virus system have a common antigen. A chemically induced tumour at a site in the body has a specific antigen as exemplified by methylcholanthrene following its injections at different sites.

Anti tumour effector mechanisms (both cell mediated and humoral immunity) are mechanisms in the body to destroy tumours.

These mechanisms are as follows:-

1. Cytotoxic T lymphocytes

Specifically sensitized cytotoxic T cells are known to destroy experimental tumours in the body. These cells provide protective T cell immunity to patients suffering from virus associated neoplasms.The tumour infiltrate of lymphocytes in stained section is directed against Tcells defined tumour antigens and do not indicate mere presence of inflammatory cells in tumours or in the area adjacent to tumour cells.

Natural killer cells (NK cells)

These NK cells are capable to kill tumour cells. The cytokine IL-2 stimulates NK cells to kill tumour cells and this process is the first line of defence against tumours with low level of class I MHC molecules in cells . NK cells also participate in antibody dependent cell mediated cytotoxicity (ADCC).

Macrophages

Macrophages, T cells and natural killer (NK) cells collaborate or act together to destroy the tumour cells. IFN-α, which is produced by activated T cells, activates the macrophages. Then, these activated macrophages kill the tumour cells. These cells produce reactive oxygen free radicals and TNF-α (tumour necrosis factor-α) to destroy the neoplastic cells. TNF-α increases the cytotoxic potential of macrophages to induce lytic effects against tumours. Humoral mechanism includes activation of complement and induction of ADCC by NK cells to destroy tumour cells.

Some mechanisms adopted by tumours to escape and evade the host immune system.

These are as follows:-

1. Immunosuppression

Certain oncogenic effectors like ionizing radiation and chemicals suppress the immune response of the host. Tumours and their products e.g. TGF- α exert immunosuppressive effects on the host

2. Apoptosis of cytotoxic T cells

This is seen in cases of melanomas and cellular carcinomas which express Fas ligands. Such tumours kill Fas ligands expressing T lymphocytes.Thus, tumour specific Tcells are destroyed in the body. In other words, there is an apoptosis of cytotoxic T cells.

3. T cells are sensitised by costimulatory molecules and peptides (antigens) bound to antigenic clefts in class I MHC molecules. These class I molecules transport the peptides derived from tumour cells to their surface for sensitisation or recognitaton of T cells. MHC class I molecules do not elaborate costimulatory molecules and thus, evade destruction by T cells. MHC molecule deficient tumour cells also escape T cell recognition. Class I MHC molecules inhibit cytotoxic effect of NK cells by interacting with receptors expressed on natural killer cells.

Some important facts are described in this paragraph in relation to immunity of tumours. Immunity against tumours induced by certain factors e.g. DNA and retroviruses (i.e some RNA viruses) is related to roles of cells like T-cells as well as B-cells. B-cells, if programmed, may produce some proteins (immune bodies) which bind to foreign proteins or antigens on the causative factors like viruses. Cancer cells make proteins which are different from those produce by normal cells. Cancer cells produced by certain viruses possess some proteins (say, viral peptides) which are carried to the surface of neoplastic cells by MHC-class 1 molecules (major histocompatibility complex class 1 molecules) attached in their antigen binding clefts . Such antigenic viral peptides called tumours specific antigens (i.e. TSAs) evoke a cytotoxic T-cell respone through the sensitized CD8+lymphocytes. These cytotoxic T lymphocytes cause destruction of the tumour cells. NK (natural killer cells) also killed the tumour cells both the humoral i.e. antibodies dependent cellular cytotoxicity- ADCC and cellular immune responses work together to destroy the cancer cells. Oxygen free radicals (super oxide anion radical O_2^- hydrogen peroxide $H_20._2$ and hydroxyl ion $OH^.$ and certain cytokines (e.g. TNFα) secreted by macrophages cause lysis of neoplastic cells. Little information on such similar immunotherapeutic major in respect of malignant tumours (e.g. sarcomas) produce by metazoan or helminthic parasites (*Spirocerca lupi* in the oesophagus in dogs and *Cysticercus fasciolaris,* a hepatic bladder worm in rats). In short, tumours show immunogenicity and immunotherapy is directed against specific antigens expressed on the surface of tumour cells. Infiltration of lymphocytes within in the vicinity of certain tumours e.g. squamous cells carcinoma indicates a cellular immune response. Inhibitors due to autoimmunity influence the immune system to clear the tumours cells.

Epithelial and non-epithelial tumours

Tumours arising from different kinds of parent tissues vary from each other to a great extent in their microscopic appearance and character (like benign or malignant) and these can be properly identified on the basis of the distinguishing features for prompt treatment. Below are the broad groups of tumours:

A. Epithelial tumours

B. Non-epithelial tumours

A. Epithelial tumours

Epithelial tumours arise from epithelial tissues and can be divided into two groups, namely,

1. Benign tumours
2. Malignant tumours

For example, papilloma and adenoma are benign tumours where as carcinoma and adenocarcinoma are malignant tumours.

(A) BENIGN TUMOURS

1. Papilloma 2. Adenoma

(B) MALIGANT TUMOURS

1. Carcinoma 2 Adenocarcinoma

Papilloma

This tumour is quite common in domestic animals. The common sites are abdominal wall, udder and neck in cattle, prepuce, penis and eyes in the horse and mouth and anal region in the dogs. It grows from, the epithelium of skin and mucous membrane and polypoid in the alimentary canal. Buccal warts in dogs and common warts in cattle are known examples of the papillomas. Stratified squmous epithelium in larynx or elsewhere can give rise to a papilloma which may be single or multiple with warty projections. The tumour grows as finger like projections from the skin surface or mucous membrane. These growths may resemble cauliflower or exist as small rounded projections.

Microscopic appearance

It consists of a core of connective tissue with a cap like covering of epithelial tissue. Stroma contains blood vessels in order to nourish the growing epithelium and a proportionate link is seen between the growth of epithelium and connective tissue stroma of the tumour. The epithelium of the tumour may be squamous, columnar or transitional in shape according to the tissue of the origin of the tumour. There are no hairs or normal glands in the tumour.

Polyps

These are not neoplasms but represent the inflammatory growths in some hollow organs. These growths have been found in thje lower bowel and also noticed in the nasal passages of cattle and horses.

Microscpically, it consists of loosely arranged fibrous or myxomatous tissue with a covering of epithelium. Inflammatory cells of various kinds are found in such growths.

Adenoma

It is a tumour arising from the glandular tissues of skin and mucous membranes, liver, kidneys, and mammary glands etc. It is a benign tumour with a capsule. This tumour forms no metastasis or infiltrative growth and may grow to a size of golf or cricket ball in an organ. It has been found in stomach, intestines, prostate, thyroid and ovary etc. The neoplastic growths arising from the mucous membranes are polypoid in appearance. Adenoma can arise from circumanal glands in dogs. Adenoma has been found in cattle and sheep and also in the thyroid, kidney and bladder of horse (Fig. 31).

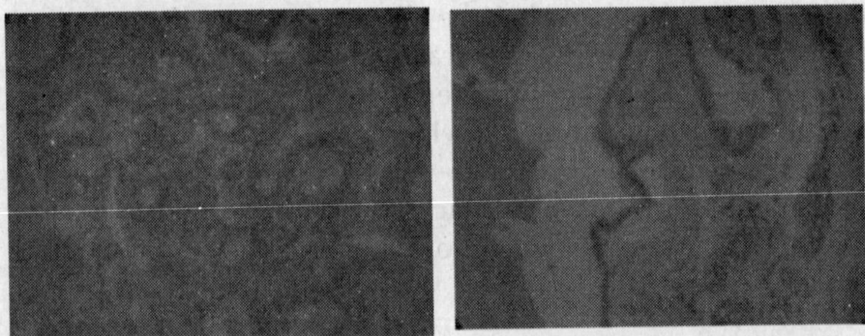

Fig. 31 Adenoma, thyroid gland dog. Note the colloid in the thyroid follicles of the thyroid gland H & E x 400

Fig. 32 Papillomatous adeno carcinoma,horse. Note the papillary proliferation into the glandular lumen H & E x400.

Microscopic appearance

It consists of glandular epithelium and connective tissue and resembles the normal gland in structure. The epithelium of the adenoma is regularly arranged and a basement membrane demarcates it from the stroma of the tumour. There is no proper development of the ducts in the adenomas and cystic dilatation of tubules and acni often occurs in such tumours. The cysts in these structures contain the inspissated secretions of the sebaceous glands, presence of several layers of glandular epithelium intstead of single layer and branching of the papillary projections of the glandular show the tendency of the neoplasms towards malignancy.

Malignant epithelial tumour (Carcinoma)

This tumour grows from epithelium in different structures like skin, mucous membranes and glands etc. The epithelium in the eye structure of the Hereford breed of cattle frequently shows cases of carcinoma. It is very common in cattle, horse and cat. Its occurrence is rare in young animals. The sheep, goats and pigs are food animals which are killed at a very young stage in their life for meat and do not usually reveal carcinomas in their bodies. It has been reported in the skin, tonsils, gums, anus, thyroid and mammary glands in dog, tongue and oesophagus in cats and penis, sinus of the head and kidney in the horse. This tumour forms metastasis and infiltrative growths and grows very rapidly and recurs after removal. The term cancer is a non-commital term applicable to a malignant tumour of epithelial or non-epithelial origin.

Microscopic appearance

It consists of groups of epithelial cells with a varying amount of connective tissues arranged around them. These groups of epithelial cells are cemented together and lie embedded in a connective tissue stroma. The cancer cells are seen fastened (one to the next) in the clumps of cells. These malignant cells break through the basement membrane and pentrate into the neighbouring connective tissue stroma. Burrowing columns of epithelial tissues act like foreign material which causes proliferation of the connective tissue stroma and the cancer cells often enter the lymphatics or paths of the least resistance in the organs and get lodged in the neighbouring lymph nodes which also show metastatic growths. Scirrhous carcinoma is hard in consistency due to predominant presence of connective tissue in the tumour. The encephaloid carcinoma is soft in consistency and is also rich in cells. Infiltration of lymphocytes and plasma cells may be found around the columns of the peneterating cancer cells into the substances of the affected organs. Carcinoma arising from glandular cells or ducts in the solid organs like liver usually forms an in ill-defined mass.

The chief forms of the cancers are as under :

1. Squamous cell carcinoma

2. Basal cell carcinoma (Rodent ulcer)
3. Spheroidal cell carcinoma
4. Adenocarcinoma

I. Squamous cell carcinoma

It arises from stratified squamous epithelium of skin, mucous membrane and structures or organs like tongue, mouth, tonsils, anus, penis, pharynx, oesophagus and vagina etc. Tumours of this sort also grow from the transitional epithelium of organs like bladder, ureter and renal pelvis, membrana nictitans, glans penis and prepuce in horse, skin and tonsils in the dog, tongue and oesophagus in the cat have also revealed the growth of squamous cell carcinoma. The tumours may be nodular, warty and may also ulceration.

Microsccopic appearance

Cells-nests (islands of epithelium) are formed in the squamous cell carcinoma of the skin. Basal cells are noticed in the clumps of the cancerous growth in skin but keratinized laminated structures of cells known as pearls are seen at its centre . Other cells like prickle cells etc. are also found in such cancerous growth. Lymphocytic infiltration is seen at the margin of burrowing columns of the cancerous mass. Anaplasia, lack of differentiation of the tumour cells and hyperchromatic dark staining nuclei are present in the tissues sections of the tumour. Acini may be enlarged to form large cysts like structures in a kind of tumour called cystadenocarcinoma.

Basal cell carcinoma

It originates from basal layer of the epithelium (i.e. epidermis) or from the epithelium of hair follicles and sebaceous glands and has been noticed in horses and dogs. There is no metastasis of the tumour and no recurrence after removal. It tends to ulcerate and is also called rodent ulcer.

Microscopic appearance

The neoplastic cells are smaller and rounder than those of the squamous cell carcinoma and stains deeply with basic stains.

These cells form a densely packed mass with blunt projections in the stromal tissue.

Glandular carcinoma

It arises from the glandular epithelium of the glands and can be divided into the following two types :

1. Adenocarcinoma
2. Sphreoidal cell carcinoma

This tumour has been noticed in organs like thyroid glands in the horse and dogs and mammary glands in dog, liver in ox and kidneys in horse. Stomach, intestines and bile ducts in animals have also revealed such tumours.

Adenocarcinoma

Microscopically, the malignant cells arise from the glandular epithelium of the glands of which glandular arrangement is imperfectly preserved. The tubules of the acini in the neoplastic mass are destroyed and irregularly arranged and several layers of lining cells can be seen in the acini. Basement membrane of the acini cannot be visible and cancerous cells are seen infiltrating the surrounding tissues. The tumour cells form columns of closely packed polyhedral cells in adenocarcinoma of liver and resembles the liver cells. The stroma in the hepatic cancer may be scanty or extensive and the neoplastic cells differ from the tissue of origin in morphological appearance. Extensive papillary projections covered by columnar or cuboidal epithelial cells are found in papilliferous adencocarcinoma. (Fig.32)

Spheroidal cell carcinoma

The neoplastic cells of such tumour show masses of alveoli of cells of irregular shapes and sizes which are surrounded by fibrous tissue stroma. The tumour cells are spheroidal or polygonal in shape due to pressure of the burrowing cancer cells. The stroma is dense in the scirrhous type of spheroidal cell carcinoma which has also less numerous presence of the tumour cells than encephaloid type. The encephaloid type grows rapidly and is poor in stromal content and has also larger type cells than the scirrhous carcinoma.

B. Non-Epithelial tumours

Fibroma; Fibrosarcoma

It is a benign tumour arising from the connective tissue of skin, periosteum, fascia and nerve sheaths, uterus and vagina etc. It can occur in any part of the body possessing connective tissues Fibroma is a benign tumour but some may recur after removal. It has two varieties.

(i) Hard fibroma (Fibroma durum) – Hard in consistency

(ii) Soft fibroma (Fibroma molle)- Soft in consistency

Gross appearance

These tumours are rounded or lobulated and have capsules and can grow up to a size of a foot ball. Hard forms (fibroma durum) are usually larger than the soft forms. The cut surfaces of the hard tumours show irregular wavy bands of white or silky fibrous tissue. Soft tumours present pink surface due to rich supply of blood.

Microcscopic appearance

Bundles of white fibres containing vessels in varying proportions are found in the cut sections of tumours and the type cell of the tumour is a fusiform connective tissue cell. Collagenous fibres produced by the tumour cells separate them from each other. Blood vessels and little vascular fibrous tissue represent the stroma of the tumour. Fibroma durum reveals predominance of fibrous tissue whereas in fibroma molle, cell bodies and nuclei predominate with presence of a few fibres. A fibrosarcoma shows high degree of cellularity and hyperchromatic staining of nuclei and nuclei may be plump, rounded or even stellate.

Some authors have divided malignant connective tumours (sarcomas) into large and small types on the basis of the shapes and sizes of neoplastic cells which may be either round or spindle shaped in appearance. The type cell of the round cell sarcoma may be either small or large in size and the same pattern is also applicable to the spindle cell sarcoma. A classification of sarcoma is given below :

SARCOMA

1. Round cell sarcoma 2. Spindle cell sarcoma

A. Small type B. Large type C. Small type D. Large type

Presence of both kinds of these cells (round or spindle type) characterizes a type called mixed sarcoma (i.e. a mixed tumour)

Sarcoma

Sarcoma is a malignant connective tissue. The neoplastic cells resemble embryonic connective tissue cells and possess delicate intercellular stroma. The tumour multiplies in all directions, destroys and replaces the tissue of origin. Areas of haemorhage and necrosis are seen in these tumours which form metastases chiefly by entering the blood stream. There is a rapid growth shown by such tumour with marked local infiltration. The tumour recurs after removal from a few remaining living neoplastic cells. Fibrosarcoma is made up of atypical fibrocytes separated by collagen which is made up of typical fibrocytes separated by collagen which is produced by these cells. The neoplastic cells may be bizarre in forms with several mitotic figures.

The intercellular stroma in a highly anaplastic tumour is scanty, illformed somewhat and seems mucoid in nature.

Round cell sarcoma

Gross appearance

These small and large types of sarcomas can be found in subcutaneous and submucous connective tissue, periosteum and intermuscular septa etc. These tumours may be large or very soft in consistency and their cut surfaces are haemorrhagic and matastasis often occurs via the blood stream. The small round cell sarcoma is the commonest and a most dangerous variety which is very soft and grows very fast with a poor prognosis. (Fig. 33).

Fig. 33 Round cell sarcoma, horse. Note the numerous round cells. H & E x 400.

Microscopic appearance

Haemorrhages are very common in round cell sarcoma which is highly cellular with scant intercellular stroma. This tumour has very rudimentary blood vessels like capillaries consisting of a single layer of endothelial cells. The large round cell sarcoma is less common and less malignant in comparison to the small variety. The small round cell of the sarcoma looks like an early embryonic fibroblasts (Fig. 33). Blood vessels are minute and numerous and its intercellulalr stroma is scanty.

Spindle cell sarcoma

This tumour is found in the subcutaneous tissue, fascia and periosteum. It is firmer than round cell sarcoma and is not so much matastatic. Its growth is slower than that of round cell variety.

Microscpic appearance

The type cell of the tumour is fusiform or spindle shaped cell with tapering ends. It looks like fibroblasts in the granulation tissue. The stroma in the tumour is intercellular and usually scanty.

Mixed cell sarcoma

Mixed cell sarcoma consists of different kinds of cells, which may be small, large, round or fusiform in shape i.e. such tumours represent a mixed state of different neoplastic cells.

Myxoma, Myxosarcoma

It arises from subcutaneous and submucous tissues. It is also found in the heart and bones of the body and has been noticed in the heart of cattle and parotid gland of cattle and horse. This tumour is benign in nature and grows slowly. It looks like Wharton's jelly and is soft in consistency and translucent or brown in colour due to haemorrhages.

Microscopic appearance

There are stellate cells in the neoplastic matrix with long interlacing processes. The translucent homogenous matrix of the tumour is of mucoid nature.

Lipoma; Liposarcoma

Liposarcoma shows the general features of malignancy and lipoma grows slowly and does not recur after its removal. Lipoma consists of adipose tissue and arises from the subcutaneous tissue and is innocent in nature and can grow in circumscribed or diffused way. It may be pedunculated with a fibrous capsule as noticed in the mesentery of horses and dogs. In liposarcomas, there are areas of anaplastic fibrous tissue in company with adipose tissue. Rudimentry fat cell vacuoles can also be seen.

Chondroma; Chondrosarcoma

Chondroma occurs in bone and can grow to form cartilage. It is noticed as an irregular nodular or lobulated tumour and has been reported in the mammary glands of bitches, testicles, sub maxillary and parotid glands of animals. The tumour is benign in nature and grows slowly.

Microscopic appearance

It consists of islands of hyaline cartilage surrounded by vascular fibrous tissue. The tumour is nourished by blood vessels existing in the stroma. Trabeculi which divide the tumour into lobules arise from the capsule of the tumour. Chondrosarcoma has general histological characteristics of fibrosarcoma and the neoplastic cells are like anaplastic fibroblasts.

Mesothelioma

This tumour is not a common type and may grow from mesothelium of serous surfaces in pleura and peritoneum of cattle and occurs in the form of sheet with papillomatous growths. It rarely occurs after removal, grows slowly and sometimes forms matastases.

Microscopic appearance

It consists of mesothelial cells. The tumour cells are epithelioid in nature and are grouped together into islands or alveoli by bands of fibrous tissue. The tumour is nourished by blood vessels existing in the stroma. Trabeculi which divide the tumour into lobules, arise from the capsule of the tumour, Chondrosarcoma has general histological characteristics of fibrosarcoma and the neoplastic cells are like anaplastic fibroblasts. There is no intercellular stroma in the tumour which looks like carcinoma in the tissue sections.

Osteoma, Osteosarcoma

Osteoma arises from the bones and has been reported from the mammary glands of bitches, and testicles of horses. Peritoneum and mesentery in cattle and dogs are the common sites of such tumours. Compact osteoma and spongy osteoma are its two varieties. This tumour is benign in nature. Compact osteomas which appear as small, round and hard like ivory, have been found in the bones of frontal sinus, orbit and maxillae and scapulae. Spongy osteoma looks like cauliflower and can seen at the ends of long bones such as femur, tibia and humerus. It can also arise from epiphyseal cartilages.

Microscopic appearance

It consists of irregular islands or spicules surrounded by proliferative connective tissue. Osteosarcoma has general features of malignancy. Anaplasia is apparent from lack of resemblance of tumours cells to the tissue or cells from which they arise.

Angioma

It represents neoplasms growing from blood or lymph vessels and consists of dilated or proliferated blood or lymph vessels. The growth is slow, multiple and benign in nature.

It has two forms :

1. Haemangioma

It is divided into two types, namely, (i) Cavernous haemangioma and (ii) Plexiform haemangioma.

Haemangioma, Haemangioendothelioma.

Gross appearance

Haemangiomas appear as irregularly shaped deep congested areas. When cut, blood flows from the dilated blood vessels. Plexiform haemangiomas frequently occur in skin and subcutaneous tissue and is bluish red in colour.

Microscopic appearance

Cavernous haemangioma consists of dilated vessels or intercommunicating spaces lined by endothelial cells. These spaces are separated from each other by fibrous septa.

There are dilated or proliferated arteries or veins or capillaries in plexiform haemangiomas. Microscopically, the plexiform haemangioma has dilated or proliferated arterises. Haemangioendothelioma has malingnant features. Its neoplastic cells are ovoid cells with hyperchromatic nuclei and scanty cytoplasm. The distinguishing feature of this tumour is the ability of neoplastic cells to form small capillaries. Such tumours bleed exceedingly.

2. Lymphangioma, Lymphangioendothelioma

Lymphangioma consists of dilated lymphatics and cavernous spaces. These growths form cysts in the skin and subcutaneous tissues. The cavities in the lymphangioma are of various shapes lined by endothelial cells and separated from each other by connective tissue septa.

Both tumours are comparable to haemangiomas.

Rhabdomyoma, Rhabdomysarcoma.

Rhabdomyoma is a congential tumour which rarely occurs and has been noticed in the myocardium of cow, pig and muscle of shoulder in horse and sheep. It arises from striated muscles in the organs consisting of such tissues. Rhabdomyosarcoma has been reported in skeletal, muscle, tongue, pharynx, myocardium and urinary bladder etc. In myocardium, the tumour appears as grayish nodules.

Microscopic appearance

It arises from striated muscles (skeletal or cardiac) and the neoplastic cells are imperfectly developed and rarely show complete striations. In highly undifferentiated rhabdomyosarcoma, pleomorphic cells, often with bizarre giant and multiple nuclei are present. The tumour cells are highly disorganized and anaplastic with large nuclear and cytoplasmic extensions known as strap cells.

Leiomyoma, Leiomyosarcoma

The tumour may be round, firm, circumscribed and pedunculated. Growths may be single or multiple. It grows from nonstriated muscle tissue and has been noticed in the organs like uterus and vagina in the mare, cow and sow, intestine and stomach in the horse and urinary bladder in the carnivores. It does not recur after removal from the body. There is a predominant presence of fibrous tissue in the uterine fibroid (a kind of leiomyoma rich in fibrous tissue).

Microscopic appearance

It consists of smooth muscle cells with rod shaped nuclei with rounded ends. The muscle cells have tendency to lie parallel with each other. Malignant tumours have bizarre and hyperchromatic nuclei and several mitotic figures.

Cholesteatoma

It consists of cyst like structures having an epithelial capsule of which horny scales are desquamated to form soft putty like pasty contents of the cyst. No metastasis of such tumour occurs. The tumour resembles a cluster of pearls due to the presence of cholesterol and other lipids in the epithelial appearing cells forming the walls of the growth. Such tumours have been noticed in the brain of horses.

Neuroma

It refers to a tumour arising from the nerve cells and fibres. Tumours of nervous system in animals are not of rare occurrence but these tumours frequently occur in animals like brachycephalic breed of dogs. Tissue of origin is the basis of classifying the tumours in the nervous system. Nerve cells, glia, neuroepithelium, meninges and blood vessels in the central nervous system give rise to tumours as exemplified by glial tumours i.e. astrocytoma, oligodendroglioma, glioblastoma, spongioblastoma and medulloblastoma etc. Gliomas arising from the neuroglia in the brain are soft in consistency and prone to haemorrhages. The tumours with malignant nature show the invasive property.

Neurilemmoma (Schwannoma)

It is a benign tumour arising from Schwann cells surrounding the axons of the peripheral nerves. It is composed of elongated spindle shaped cells arranged in a distinct pattern. The cells form interlacing fascicles with clumps of cells lying in windrows. This palisading (stakes formation) of the nuclei is the main undulating pattern and may form tight whorls. Such tumours occur at a plexus or ganglion or along course of a nerve and are common in cattle.

Meningioma

This tumour forms a more or less spherical subdural mass which compresses the underlying brain or cord. It has been reported in the cat, dog, cattle and horses and arises from the cranial and spinal meninges (i.e. from mesothelial cells). These mesothelial cells form the surfaces of such meninges. The neoplastic cells are more plump and the nuclei are more rounded than in the neurofibroma and the cells have tendency to make whorled arrangement. A compressed hyaline or calcified body which is visible grossly like a grain of sand is noticed at the centre of the whorls of the cells. Such structure of meningioma is called a psammoma (from psammos, meaning sand) and these granules are called brain sands.

Neurofibroma, Neurofibrosarcoma, (Neursarcoma)

These tumours arise from Schwann cells and have been reported in dog and cattle. Such tumours are also belived to grow from the fibroblasts of the perineurium. Microscopically, the benign type is composed of spindle shaped cells arranged in circular whorls or wavy fasiculi.

Microscopically, neurofibrosarcoma may closely resemble a fibrosarcoma and its cells form plump spindle shaped cells or retain a whorled or curly arrangement.

Osteoclastoma

It grows chiefly in connection with bones, which undergoes absorption leaving behind a mere shell of the bone. It breaks into pieces giving an egg cracking sound. Tumour has been noticed in bones like tibia, femur and lower jaw in dogs and horses. Osteoclastoma on the lower jaw in horse is called **epulis,** It rarely recurs after removal and seldom shows metastasis and grows very slowly.

Venereal granuloma (Infective sarcoma)

This tumour grows from connective tissue and is frequently found in the genital passages and organs of the dogs. It is a transmissible tumour and spreads to healthy dogs by coitus and is frequently noticed in the vagina and penis of dogs. If the tumour cells are injected in the subcuits of healthy dogs, infective granuloma appears in them. Spontaneous disappearance of the tumour renders the dog resistant to further infection as noticed in bacterial and viral diseases. One can notice often venereal granulomas in the indigenous street bitches or dogs transmitting the infection to others by coitus.

Microscopic appearance

Microscopically, it resembles sarcoma and consists of large round cells and bears resemblance to lymphoid cells. The cells and nuclei vary in sizes and mitotic figures are numerous.

Teratoma

Some misplaced or isolated embryonic tissues in the body of an organism multiply to develop into a peculiar tumour consisting of different kinds of cells. These tumours are called teratomas. Below are its two forms.

1. Dermoid cyst
2. Dentigerous cyst

1. Dermoid cysts

These are rounded or egg shaped cystic structures with a thick wall. Yellowish white greasy material containing entangled mass of hairs etc. are present inside the tumours. It has been noticed is subcuits, ovary and testicle. Horses and cattle have shown such tumours.

2. Dentigerous cyst

It is found in the form of a cyst containing one or more imperfectly formed teeth. It is often lodged between the temporal bones of the skull and has been noticed in horses.

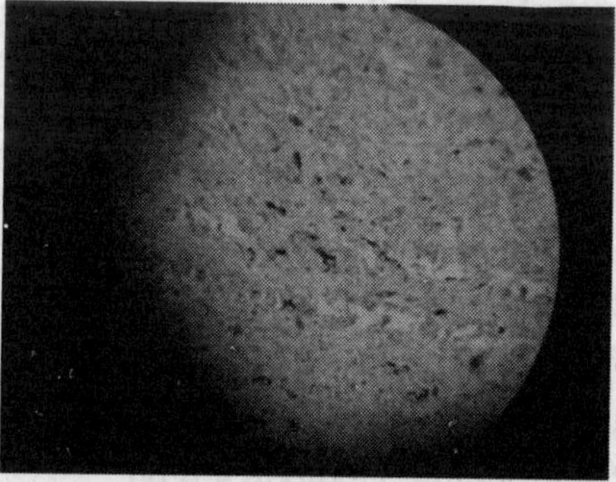

Fig: 34 Melanosarcoma, cattle. Note the deposits of melanin pigments and small round or inregular shaped cells.

Melanoma

These tumours orginate from melanin forming cells called melanoblasts and tumours may be benign or malignant and are common in perianal region and perineal skin and occur in cattle, pig and grey horses. Grossly, the tumour is deep black in colour which is visible on slicing the tumours. Microscopically, the malignant tumours reveal mitotic figures. In tissue sections, the malignant cells of melanomas are filled with melanin (i.e. a brown pigment, (Fig. 34) but in an amelanotic melanoma, no melanin is present in the cells. Melanin in urine (melanuria) is noticed in the patients of melanomas e.g. cattle. Melanomas are temporarily benign and, later, turn to be malignant. The general features of malignancy are also applicable to them and metastases are noticed in organs like lungs, spleen and other internal organs.

Malignant lymphoma (Lymphosarcoma)

It is a neoplasm of blood forming tissues which consists of lymphoid cells. Lymph nodes in the different parts of the body are enlarged in the affected animals. The neoplasm is multicentric and arises from a generalized proliferation of normal lymphocytes in an uncontrolled way. Such neoplasms are always malignant and invade almost all organs of the body and produce metastases and infiltration of neoplastic cells is found among the normal histological structures. Lymphoid cells are present in the intertubular stoma in the kidneys and among the muscle fibres in the heart. It occurs in bovines and goats which seem to be susceptible. Metastases can be found in liver, walls of reticulum, abomasum, uterus and ureter. In dogs, it is found in the lymphoid and parenchymatous organs and roof of the dog's mouth. The main classes of malignant lymphomas are as below:

1. Malignant lymphoma, lymphocytic type
2. Malignant lymphoma, histocytic type
3. Malignant lymphoma, stem cell type
4. Malignant lymphoma, Hodgkin's type

The leucocyte count in the blood of the affected animal goes very high (e.g. between 30000 and 200000) due to an increase in the number of leucocytes (lymphocytes). This condition is called lymphatic or lymphoid leukaemia. The malignant lymphoma accompanied by leukaemia is known as leukaemic type and the type without leukaemia is called aleukaemic. It may be leukaemic at one time and aleukaemic at the another. Microscopically, the neoplasm consists of uniformly large cells with several typical mitotic figures. The capsule of the affected lymph node is extensively infiltrated with lymphoid cells. Neoplastic invasions of organs such as heart, stomach, kidneys or uterus may also be present.

Chordoma

It has been reported in animals and is a rare neoplasm of humans located at the upper and lower extremities of the vertebral column. It originates from **echordoses** which are persisting remants of the foetal notochord. Grossly, it looks like a gelatinous semitransparent tissue. Microscopically, it consists of large rounded cells with central nuclei and clear mucoid appearing cytoplasm. Numerous vacuoles are seen in the cytoplasm and extracellular material is also present in the tissue sections. The neoplastic cells form clumps which may be surrounded by a noncellular matrix.

Sertoli cell tumour

This tumour of feminizing nature orginates from the Sertoli cells of the seminiferous tubules. Grossly, it consists of one or more nodules greatly distending the tunica albuginia. When cut, nodules present grey or light yellow surface. Microscopically, it consists of anaplastic areas showing solid masses of round or polyhedral cells with central nuclei in the tissue sections.

Seminiferous tubules of normal shapes and sizes can be found and the lining cells of the tubules may show one or two irregular layers of tall Sertoli cells. These tumours have been noticed in dogs and produce feminizing hormone which causes enlargement of the mammary glands and thinness of the hairs of the belly. Other dogs are sexually attracted to such males. A state of hyperoestrogenism is produced in the affected dogs.

Interstital or Leydig cell tumour

It occurs in dogs and arises from the interstitial or androgen secreting cells of the testes. Grossly, the tumour has yellow cut surface. Fibrous trabeculae divide the tumours into various compartments consisiting of cells which are large and have more eosinophilic cytoplasm. Lipids are formed in the cytoplasm of such cells and impart foamy character to the cells.

Seminoma

It frequently occurs in the testicles of old dogs and arises from the germinal epithelium of the seminiferous tubules. Grossly, the cut surfaces of the tumours are soft, white to pale grey and bulge above the adjacent normal tissue. Microscopically, it consists of large polyhedral cells with prominent centrally located round nuclei. Conspicuous characteristic bodies are formed in the nuclei of neoplastic cells. Neoplastic cells are divided into compartments by fibrous trabeculae and also show mitotic figures.

Arrhenoblastoma

It is noticed in bitches and is a masculinizing tumour arising in connection with retae ovari (a female counterpart of the testes). The tumour consists of anaplastic cells and resembles the Sretoli cell tumours. It has also been reported in bovines and neoplastic cells are polyhedral or fusiform in shapes and masculinizing features may be noticed with females.

Hepatocellular carcinoma

The neoplastic cells resemble the liver cells, and show varying degrees of differentiation. The cells are arranged in trabecular or acinar patterns or can exist as a mixture of both patterns and invade the adjacent liver paremchyma. Mitoses and enlarged nuclei are frequently found. Cystic structures are found in the tumours and the neoplastic cells metastasise to the hepatic lymph nodes.

Cholangiocarcinoma (Interahepatic bileduct carcinoma)

It is a malignant tumour of which neoplastic cells resemble those of the intrahepatic bile ducts. Metastasis can spread to the liver, serosa and lungs of the patients. The neoplastic cells form small acini and they may be columnar or cuboidal in shape and may possess clear or granular cytoplasm. The tumour cells may also form solid cords with rare formation of lumina. It occurs in deers.

Liver cell adenoma (Hepatocellular adenoma)

It has been noticed in domestic animals and is a sharply circumscribed and usually found singly as a tan coloured growth. It compresses but does not invade the liver parenchyma. It forms no metastasis. The neoplastic cells resemble normal liver cells, and are arranged in trabecular or acinar pattern. No normal resemblance to hepatocytes is noticed in the neoplastic mass of tissue. The malignant type has highly anaplastic cells.

Recurrence

A tumour can recur at a place or in its vicinity after removal and recurrence has been noticed in cases of the malignant tumours. Recurrence is seen many years after its removal but the upper limit is taken to be 3 years for the risk of another neoplastic growth. Some times, an entirely new tumour grows after several years at the site of old tumour removed surgically in organisms.

Effects of tumours

These are as under :

1. Occurrence of cachexia (wasting), anemia, and inanition because of toxaemia arising from degenerating or necrotic neoplastic masses and behaviour of tumours as parasites in the body of host causing morbidity and mortality in the body are pathologic features of tumours. Cachexia which is a condition of wasting, weakness and anaemia is an important effect seen in cancers. Cytokines secreted by tumours are factors to produce cachexia in pataients. TNF-α

(a cytokine) reduces the appetite of cancer cases. Patients loose fat and muscle in their bodies.

2. Stenosis or obstruction in any tube like organ or and narrowing of the lumen of digestive tract. Serious problems are noticed in the patients because of location of tumours in vital areas or imfringment on the adjacent tissues or organs.
3. Ulceration of the tumours, metastasis and invasion are important effects shown by tumours. Bleeding from the ulcerated surfaces of organs e.g. stomach, vagina or intestine and infection of neoplastic lesion enhance severity of the disease. Invading neoplastic cells destroy blood and lymph vessels in the adjacent tissues.
4. Toxaemia due to breakdown products of tumours and secondary metastasis due to erosions in the blood vessels causing spread of infiltrating neoplastic cells.
5. Defence to tumour is nonspecific and involution or spontaneous regression of tumours can be seen. It is more often noticed in animals than in humans.
6. Excessive production of hormones. Hyercalcaemia is seen in cases of parathyroid adenoma. Increased functional activity in certain organs e.g. endocrine glands causes an increased hormonal synthesis resulting in abnormalities.

Other hormonal disturbances include hypergylcaemia in cases of tumour of pancreatic isletes, masculinasation in tumours of the adrenals glands, feminization and enlarged mammary glands in Sertoli cells tumours in dogs.

Degeneration in tumours

Fatty changes, hylinisation, necrosis and liquefaction necrosis can be seen in tumours. Degenerative changes occur from lack of nutrition to the growing mass of tumour when there is no balanced growth between neoplasm and its vascularisation. Myxomatous changes, amyloidosis and glycogenic infiltration can be seen in the neoplastic growths.

Chapter **9**

A Veterinary Diagnostic Approach to Some Animal Diseases

C.D.N. Singh*, Former University Professor, S.R. Singh, Shanti Singh,** Associate Professor (Gynaecology), **, S.Samataray***, University Professor, Kaushal Kumar* Assistant Professor, Imran Ali* Assistant Professor and Pankaj Kumar**** Assistant Professor cum Course Co-ordinator, Bihar Veterinary College, Patna.

The concept of establishing diagnostic laboratories in addition to provincial or range laboratories in different states is a laudable step. For the success of such endeavor, it will go a long way to save the valuable live stock and poultry etc in a state.

The subject matter in this context has been limited to point out some steps and necessary facilities for diagnosis of animal diseases in a laboratory.

A brief account of the techniques, stains and reagents used and other requirements in a laboratory.

For diseases requiring blood examinations, some important steps are as under :-

1. Collection of blood samples
2. Preparation of the blood films from an ear vein or syringe on clean and grease-free slides.

One should draw the spreader slide at an acute angle to another slide behind or back to the drop of blood and allow the blood drop to spread along the end of the spreader. Then, push the spreader slide in a single non- stop sweep to draw the blood along the slide in a thin film which is suitable for differential count of white cells and protozoal examination etc. Thick films of blood made with the help of a platinum loop on slides are suitable for detecting anthrax bacilli.

3. Performance of hemoglobin estimation and white and red cell counts with the help of haemometer and haemocytometer respectively.
4. Examination of blood films and differential white cell count.

* Department of Pathology

** Principal, Bihar Veterinary College, Patna

*** Department of Gynaecology P.M.C.H. Patna

**** Department of Parasitology

***** Department of extension

5. Calculation of sedimentation rate (Western Green method) 0.5 ml of 3.8% Sodium citrate solution in a test tube is used for 4.5 ml. of blood.
6. Clotting time

It is found out by breaking off the capillary tubes filled with blood in small pieces every half minute. One records the earliest time when clotting of the blood occurs in a capillary tube.

Commonly used stains and staining procedures

1. Gram's stain and Gram's method for staining bacteria.

Bacteria are differentiated in to gram- positive and gram - negative bacteria.

2. Ziehl-Neelsen stain and Z.N. technique.

This method is used to divide bacteria into acid-fast and non acid fast types.

3. 0.5% crystal violet solution.

It is used for staining of spirochaetes.

4. Giemsa's stain and its procedure for use.

It is used for staining protozoa and certain organisms.

5. Leishman's stain and its procedure.

It is usedful for protozoa and certain bacteria.

Some other examinations

1. Fecal examination by direct and sugar flotation methods.
2. Examination for detecting fungi e.g. ring worm.

Hairs collected from the edge of the lesions are mounted in slightly warm 20% potassium hydroxide solution. 5% KOH solution is used for examining active mange lesions. Skin scrappings are boiled for 5 minutes in 5% KOH solution before examination.

3. Examination of urine
4. Sugar, protein and acetone etc are found in urine in certain diseases.

   Bacteriological examination of urine enables one to make an etiological diagnosis for specific logical treatment.
5. Examination of blood for calcium, magnesium and acetone etc.

   Rothera's reagent helps one to estimate the ketones level in urine in some metabolic disorders.
6. Examination of intestinal contents for coccidiosis e.g. *Eimeria tenella* infection in the caecum of chickens.

   One wipes aside the intestinal contents with a blunt scalpel for making thin firms and examination of the films covered with cover slips under low and high powers of the microscope for noticing oocysts.

7. Examination of milk

It is carried out to detect pathogenic bacteria and chemical abnormalities

Postmortem examination

The object behind postmortem examination of a dead body is to find out the cause of death of an animal. A systematic postmortem examination is a must to arrive at a definite diagnosis of the diseases. One needs a postmortem set, clean slides, sterilized pipettes and 10% formalin solution in small jars.Before conduction of postmortem examination, an autopsist reads the history of the case. Special attention is given to vital organs like brain, liver, kidneys and lungs etc. The lesions developed in the organs of the animals are responsible for symptoms noticed during life. One notes down the lesions (i.e. pathoanatomical deviations) as these are encountered during autopsy of the dead bodies. Postmortem reports are not written from one's memory on some other days after postmortem examination. Only objective statements of the observations at autopsy table are recorded.

Macroscopical- (naked-eye or gross) and microscopical examinations together with microbiological, parasitological and chemical findings etc from laboratories are required to arrive at a final conclusion i.e. an ultimate diagnosis of a disease. Only observations are the back bone of pathology (an observational science). An observation should be strictly separated from an interpretation. For example, chronic passive congestion of the liver is an interpretation of the nutmeg liver. Gross lesions are described in view of their size, shape, colour, consistency and appearance of the cut surface and relationship to their surroundings. These are said to be alphabets of recording gross changes of the lesions. Life is a state of labile equilibrium between diverse, labile and potentially perpetual reactions in the cells of the living beings. As a result, a cell is the unit of life and cessation of cellular, biochemical and physiological reactions i.e. life ends in the death of an organism.

Death is marked by a gradual transition from life to death and is divided into two important kinds, namely, (i) clinical death and (ii) somatic death. Clinical death means a stoppage of heart and lungs and loss of consciousness in an animal. Somatic death means death of cells i.e. loss of all biochemical and physiological activities in them. Life still continues for some time even after corporal or clinical death in the body. Certain terms like brain death, liver death and renal death are also in use. For diagnosis of diseases, one chooses the most severe pathoanatomical lesion in a vital organ (say, lung or liver) as the cause of death. Diagnosis is divided into certain kinds e.g. etiological diagnosis and somatic (cellular) diagnosis etc. Anthrax and tuberculosis are etiological causes of deaths. Pneumonia and hepatitis are somatic or biological causes of deaths. A single infection with a pathogenic organism i.e. *Bacillus anthracis infection* causes an acute disease but mixed infectious processes (say, mixed infections) or double infections arise from an attack of the tissues of the hosts simultaneously by more than one pathogenic bacterium or virus. Viruses or parasites devitalize the tissues and pave the way for attack in the organs by another infectious agents. This is called secondary infection by bacterial invaders e.g. cocci or ***Salmonella cholerae suis*** in the organs of the host. Useful information on some different disease is given in Table 14 with suggestive figures.

Table-14 Diseases and techniques for their diagnosis in animals

Diseases	Techniques (Methods) adopted
1. Bovine abortion e.g. ***Brucella abortus*** infection	1. Examination of the aborted foetus and description of the lesions encountered in the body. 2. Cultural examination of the foetal contents. 3. Examination of the placenta. 4. Cultural examination of the milk of the suspected cases. 5. ***Brucella abortus*** agglutination test of the serum of suspected bovine for infection. 6. Mucus agglutination test to detect *Brucella abortus* and *Vibrio foetus* . 7. Brucella ring test. 8. Whey agglutination test for ***Brucella abortus.*** 9. Direct examination of vaginal discharge, uterine pus, placental fluid and sheath wash for *Trichomonas foetus*. 10. Examination of stained smears of the discharge or pus etc by Giemsa's method to detect *T.foetus*
2. Actinomycosis ***(Actinomyces bovis infection and Actinobacillus lignieresi infection)***	1. Direct examination of pus for granules 2. Cultural examination of the granules in the pus. 3. Animal (guinea pig) inoculation with fresh material or culture 4. Histopathology of the lesions in the tissues or organs.

3. Anaerobic diseases e.g. black quarter, *Clostridium* ***oedematiens*** infection and ***Cl. botulium*** infection (botulism)	1. Examination of stained direct smears by Gram's method. 2. Detection of oval, central or subterminnal spore in ***Cl. chauvoei*** (cause of BQ) and ***Cl oedematiens and Cl septique***. There is usually absence of spore in Cl. ***welchii*** 3. Subcutaneous inoculation of preformed toxin in the filtrate of suspected material or food into guinea pigs. Death is noticed in 12 hours. 4. Cultural examination of the suspected tissues e.g. muscles and organs. 5. ***Lecithinase test for Cl.oedematiens*** Presence of lecithinase is indicated by opalescence in test fluid i.e. broth culture added to 0.5 ml of emulsion i.e. an emulsion of egg yolk
4. Anthrax	1. Examination of thick blood films from ear vein by staining with 1% methylene blue or Gram's method. A blood drop is smeared on a clean glass slide with platinum loop to prepare thick smear which is preferred in suspected anthrax cases. 2. Cultural examination of material, blood and liver pieces. 3. Animal (guinea-pig) inoculation with bacterial culture by subcutaneous route . 4. Examination of decomposed tissues and food tissues etc. for ***Bacillus anthracis*** by isolation of organisms and animal inoculation. 5. Asol is test (a precipitin reaction)

	An extract of suspected materials like blood, hide, spleen and decomposed tissues and anti-anthrax hyperimmune serum react together to form a precipitate immediately. A granular purple ground work between the anthrax bacilli is created by staining of the disintegrated capsular material of ***Bacillus anthracis*** in the stained blood films by 1% ***methylene*** blue solution .
5. Glanders	1. Examination of direct smears by Gram's method to detect gram-negative rods. 2. Cultural examination of the suspected material 3. Animal (rabbit) inoculation is followed by development of orchitis in rabbits (Straus test).
6. Johne's disease	1. Direct examination of smears of moist faeces or smears from mucosa pinched from the rectum by Ziehl-Neelsen's method to notice acid -fast bacteria. 2. Cultural examination of the faeces . 3. Johnin test in suspected bovines or herd of cattle.
7. Tuberculosis Cough, low grade figure and weight loss are signs - common to T.B. in man and animals and the diagnostic methods include direct microscopy of stained smears of material or	1. Direct examination of smears from affected lymph nodes or lung nodules and milk sediment etc. by Z.N. method to see acid -fast bacteria. 2. Cultural examination of nodules. 3. Guinea- pig inoculation Two such animals are injected intramuscularly with suspected material i.e. milk to produce tuberculosis in their organs.

discharges, culture of organisms, histopathology of T.B. lesions in organs, P.C.R. (polymerase chain reaction), ELISA (enzyme linked immuno sorbent assay), X-ray and imaging etc.	4. Tuberculin test in suspected herds of cattle and buffalo to detect positive reactors of tuberculosis.
8. Swine erysipelas ***(Erysipelothrix rhusio pathiae infection)***	1. Examination of stained films or direct smears of heard blood, cardiac valvular lesions and joint fluid by Gram's method for presence of small slender Gram positive bacteria 2. Cultural examination of the suspected tissues or organs. 3. Postmortem examination of fateal cases to detect lesions in cardiac valves or joints. 4. Biological test Mice are injected with glucose broth culture to show conjunctivitis.
9. Leptospirosis ***(Leptospira spp. Infection)***	1. Direct examination -of sedimentary deposit of the blood collected at the height of the temperature to see organisms. Blood is centrifuged at 3000 r.p.m. for 5 hours . 2. Examination of urine Fresh urine is centrifuged for 2 hours at 3000 r.p.m and deposit is examinined for bacteria. 3. Examination of fresh kidney or liver tissues by Giemsa's and Levaditi's methods and dark ground microscopy .

	4. Biological examination Deposits of fresh urine centrifuged in test tubes are used as inoculum which is injected into white guinea pigs intraperitoneally. The inoculated guinea pigs are observed for three weeks and infected cases reveal icterus, and hemorrhage in lungs and muscles etc.. 5. Cultural examination Special media like Noguchi's medium is inoculated with fresh kidney and liver material or blood collected at the height of temperature of experimental animals.
10. Mastitis Some important observations in mastitis are blood clots and abnormal smell in the milk and cardinal signs of inflammation like redness, swelling, hot and painful udder etc.	1. Plating of one loopful of milk on to the Edwards medium and blood agar plates. 2. Examination of the stained films of the sediment of milk sample centrifuged at 3000 rpm for half an hour by Ziehl-Neelsen's and Gram's method of staining to detect the presence of bacteria. 3. White side test One drop of 5% Na OH is mixed with 4 or 5 drops of milk on a glass slide and the mixture is stirred with a platinum loop. Presence of granules or flecks after half minute in the

	mixture indicates mastitis in the patient. 4. The Coagulase test A single colony of the staphylococci is mixed with one drop of 1/10 dilution of rabbit plasma. Formation of clumps within a few second indicates staphyloccal mastitis. 5. Antibiotic sensitivity test A suspension bacterial isolate in 0.5 ml of broth is spread over a dried blood agar plate. A Multo disc i.e. an antibiotics disc is placed in the center of the plate for incubation. The zones of inhibition around each antibiotic disc are read to determine the sensitivity of the isolated bacteria.
Poisoning **i.** Poisoning due to fertilizers. The important signs of poisioning caused by organic phosphates (Organo phosphate compounds e.g. dichlorovos). include dyspnoea, pulmonary oedma, salivation and muscle stiffness in animals like cattle and sheep etc.	1. Estimation of the level of cholinesterases in the circulating red cells and (ii). fatal accumulation of anticholinesterases at the motar end plates indicate OP toxicity by histochemical method.
Protozoal infection 1. Trypanosomiasis ***(Trypanosoma evansi infection).*** Also called surra. in animals like cattle.	1.Stainning of thin blood films with a clear edge with diluted Giemsa's stain for 1/2 hour. The slides face down-wards in stain during staining. 2. Performance of mercuric test for Surra in camels by adding a drop of the camel's serum to 1 ml. of one in 25000 mercuric chloride solution. A ring of precipitate is seen in positive cases.

	Even formol gel test or stilbamidine test for surra is carried out by mixing 4 drops of formalin to 1 ml. of serum in a test tube. An opalescence due to high level of proteins in blood is noticed in positive cases. Stilbamidine test is a carried out to diagnose surra in cattle.
2. Babesiosis ***(Babesia bigemina infection)***	1. Examination of thin blood films with Leishman's or Giemsa's stain. Protoza are readily observed in patients showing acute signs of the disease. Pear - shaped forms (even round or irregular forms) in pairs with pointed ends contacting each other are noticed in the red cells. The parasites and their chromatin materials are stained bluish and red respectively with Giemsa's stain .
3. Theileriasis *(Theileria spp infection)* in bovines Cross bred or exotic breeds of cattle are more susceptible than indigenous cattle.	1. Examination of blood smear from the enlarged lymph nodes, spleen liver etc. Stained with Giemsa's stain to detect Koch's blue bodies. (1 to 15 microns in diameter) as bluish stained masses containing red stained chromatin dots. These bodies break down and release theilerial parasites to enter red cells which reveal red stained ovoid dots or elongated forms composed of chromatin material of ***Theileria*** *spp*

4. Anaplasmosis * (*Anaplasma marginale and Anaplasma centrale infection*) Cattle are susceptible. * For more details of information on animal diseases, see the publications 'Advanced Pathology and Treatment of Diseases of Domestic Animals (2008) and Advanced Pathology and Treatment of Diseases of Poultry by Singh, C.D.N. et al (2005), IBDCO, Khushnuma Complex basement, 7, Meera bai, Marg (behind Jawhar Bhawan, Lucknow, U.P.)	1. Examination of thin blood films stained with Giemsa's stain to detect particles of chromatin like cocci or spherical bodies in the red cells of affected cattle and sheep.

Suggestive figures (outlines) of bacteria, protozoa and eggs of parasites in some important diseases in animals

Bacterial and protozoal organisms

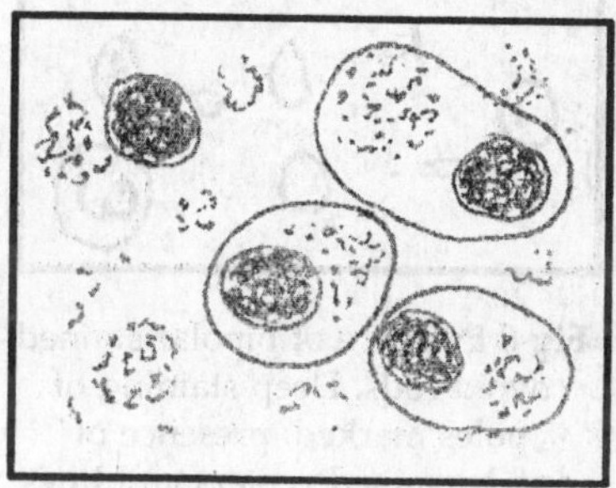

Fig-1 Note the Koch's blue bodies called schizonts as faint blue bodies containing red stained chromatin dots or particles in smears of lymph nodes stained by Leishman's method (For method see appendix)

Fig.2- Presence of slender acid-fast stained rods singly or in groups comparable to a bundle of faggots in the smears of affected lymph nodes stained by Z.N. method. Beaded forms of acid- fast rods due to un stained (colourless) spaces in the bacterial bodies. For method, see appendix.

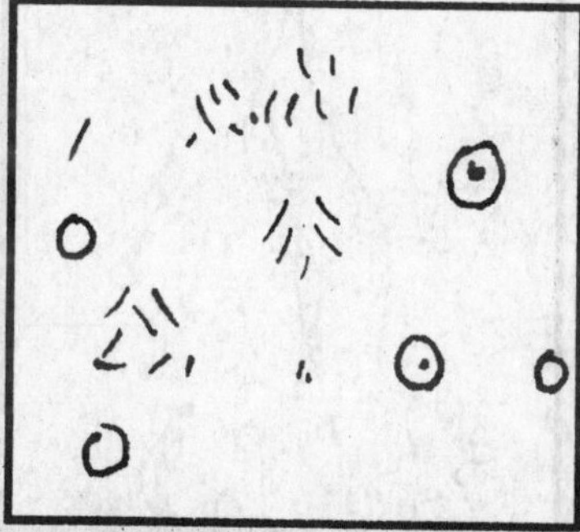

Fig-3 Note the red stained acid fast rods in smears of mesenteric lymph nodes or rectal mucosae stained by [illegible] as singles, clums or groups in the stained smears. Shorter than T.B. bacteria. See the appendix for method.

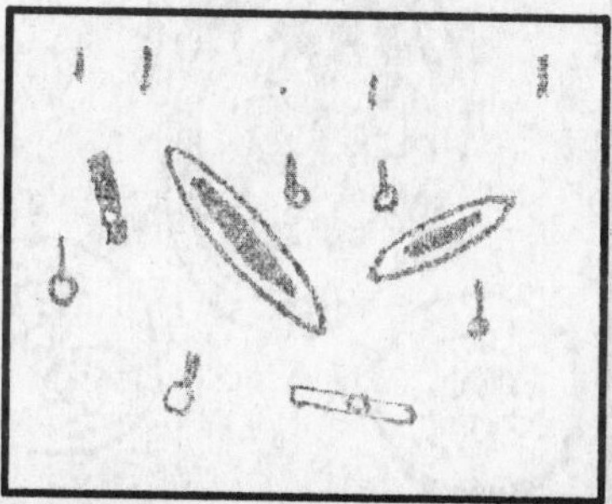

Fig-4. Gram positive navicular forms of bacteria causing black quarter ***(Cl. chauvoei as causative agent)*** Slightly curved or straight [illegible] rods in pairs are also seen. Bactarial rods bearing central or subterminal spores in the stained smears. For method see appendix.

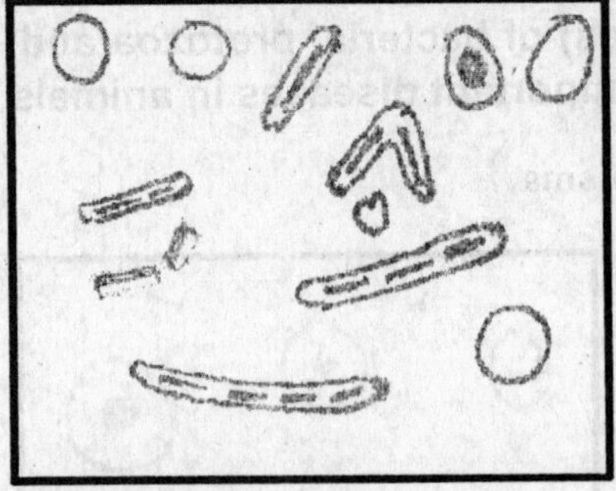

Fig-5. Note the capsules of anthrax bacilli showing square ended roads singly or in tows or threes with a comman capsule in the blood smears stained by Leishman's stained or 1% methylene blue. Very long chains of larger rods without capsules noticed in the blood or tissues smears of decomposed dead bodies For method see appendix

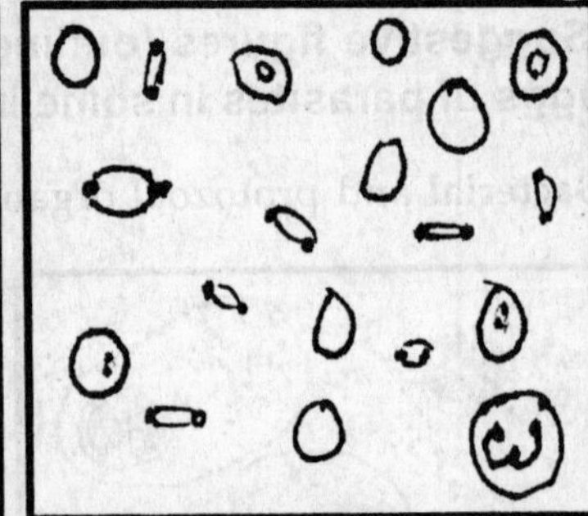

Fig-6 Presence of bipolar stained ovoid rods. Deep staining of poles marked, presence of poorly stained centers and lines linking deeply stained poles in the blood films of HS patients stained by Leishman's metnod. For method see appendix.

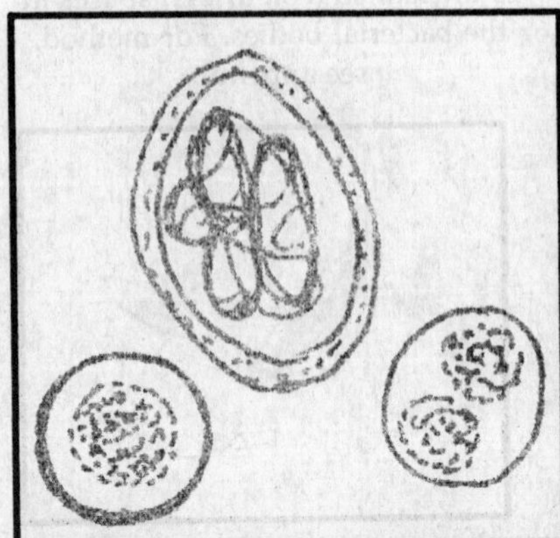

Fig-7. Note the sporulated and non -sporulated forms of coccidia. Only four sporulated oocysts containing two sporozoites seen in the preprations of faeces of *Eimeria spp.* infection

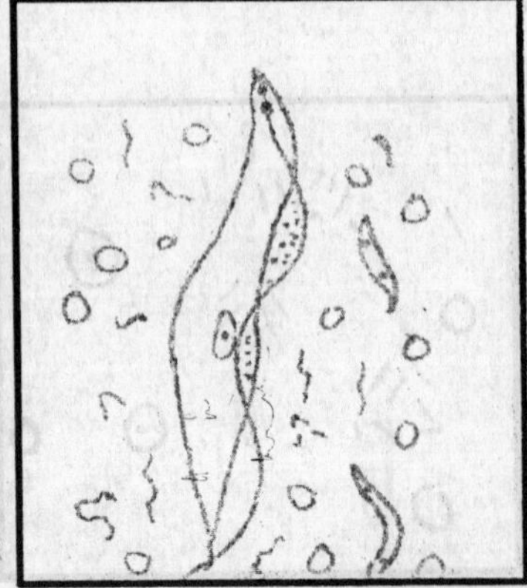

Fig-8 Presence of spindle shaped flagellated forms of Trypanosomes in the blood films stained by Leishman's method. Collection of blood for making blood smears is preferred during acute stage of Surra marked by fever

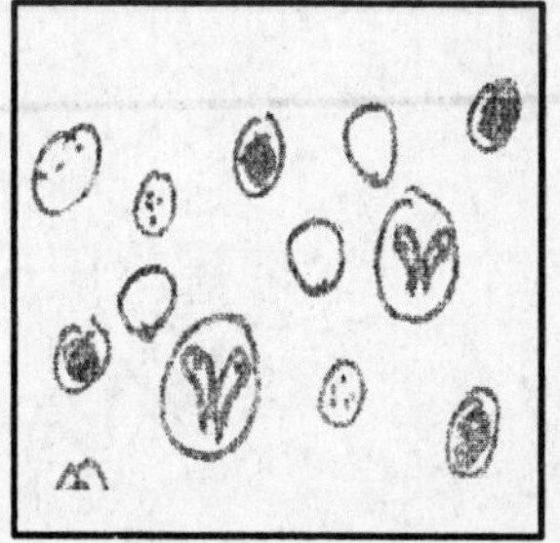

Fig-9. Pear shaped piroplasms in singles, twos or four in infected red cells. The protozoa are marked by chromatin dots, vacuoles and thin capsules in the smears stained by Leishman's method. Preparation of blood films preferred during acute disease.

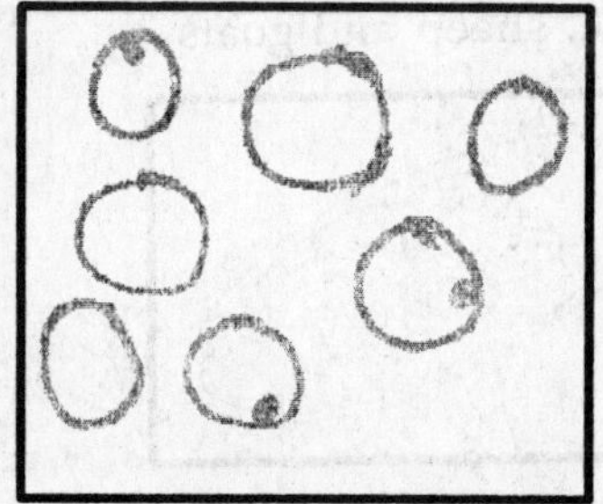

Fig-10 Note the presence of round or oval chromatin bodies (*Anaplsma marginale*) in red cells of the stained blood films by Giemsa stain.

Eggs and larvae of some parasitic helminths.

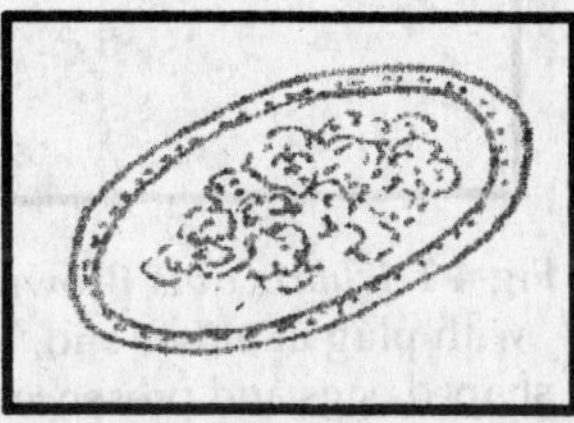

Fig-1 *Strongylus* spp and *Trichostrongylus axei* etc. Thin smooth ovoid eggs and segmenting embryo.

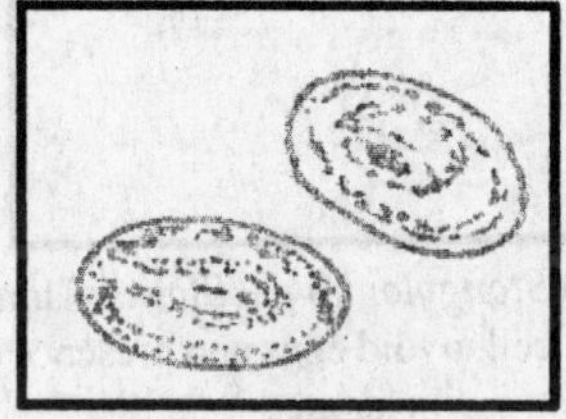

Fig-2 *Strongylus westeri* Thin smooth ovoid eggs and presence of larvae

Fig-4 *Ascaris equorum.* Unsegmented ova, globular eggs with thick brown albuminous coat .

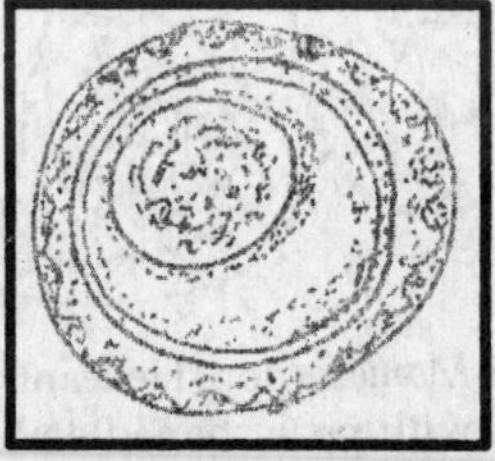

Fig-5 *Anoplocephala* spp. Thick rough coat of the shell and eggs globular flattenedd at one or more sides. Hexacanth embryo in the pear shaped embryophore

Cattle, sheep and goats

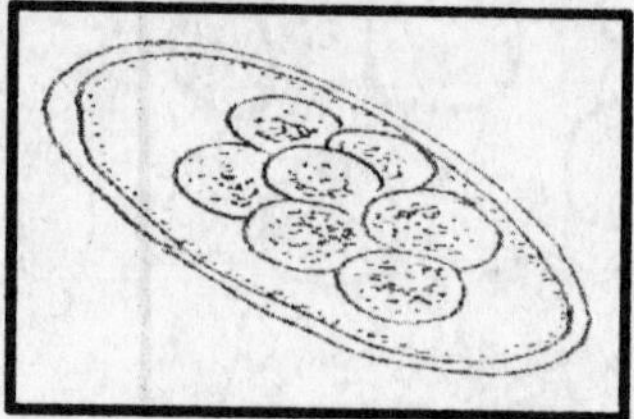

Fig-1 *Nematodirus filicollis* and *N. spathiger* Segmented embryo, smooth shell and ovoid eggs.

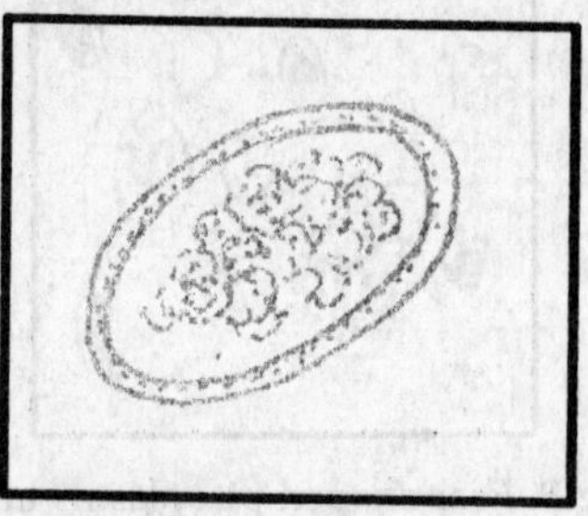

Fig-2 *Haemonchus contortus, Trichostrongylus* spp, *Bunostomum* spp. Thin smooth ovoid eggs and segmenting embryo.

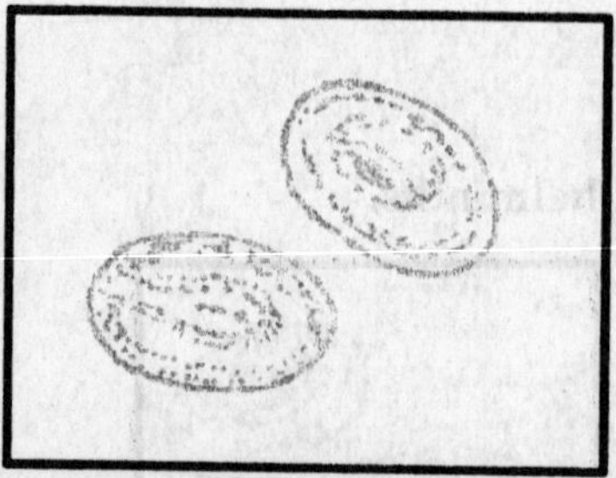

Fig-3 *Stongyloides papillosus.* Thin smooth cell, ovoid eggs and presence of larvae .

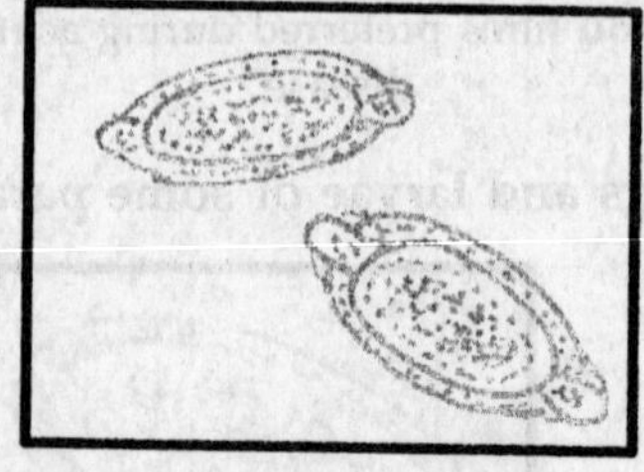

Fig-4 ***Tricluris ovis*** Brown shells with plug at either end, barrel shaped eggs and presence of the unsegmented embryo.

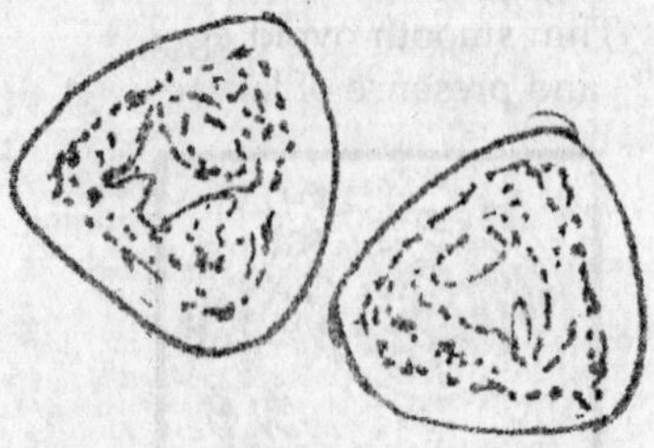

Fig-5 *Moniezia* spp. Hexcanth embryo with pear shaped embryo phore, thick tough shell and globular egg flattened on one or more sides

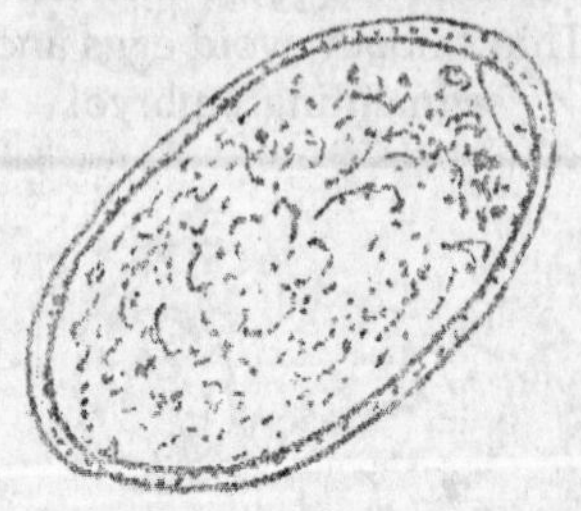

Fig-6 *Fasciola* spp Thin shell, ovoid eggs with an operculum at one end, shell field with yellow, green yolk.

Fig-7. *Dictyocaulus viviparus and D. filaria.* Larvy containing granules and button at head end of the larvae.

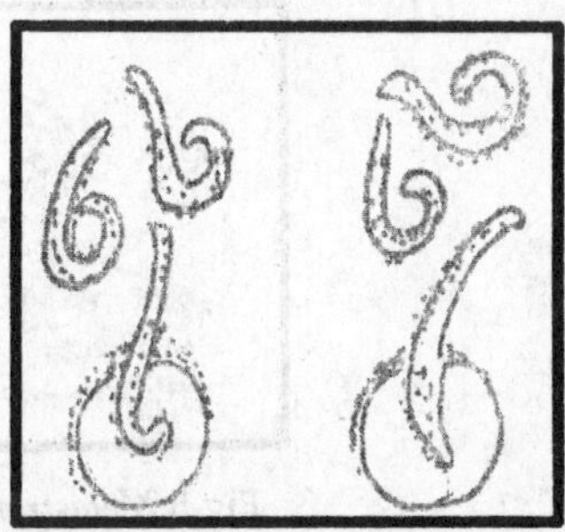

Fig-8. *Muellerius capillaris* and *Protostrongylus rufescens.* Wavy process on the tail of the larvae.

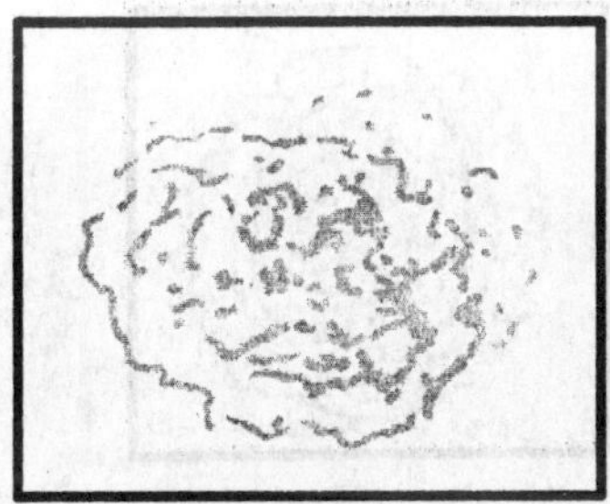

Pigs

Fig-1. *Ascaris lumbricoides* Ovoid ova, thick shell with albuminous coat and presence of unsegmented embryo.

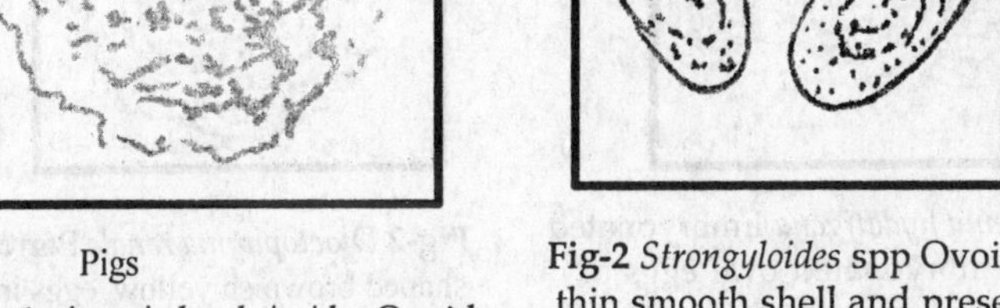

Fig-2 *Strongyloides* spp Ovoid eggs, thin smooth shell and presence of larvae.

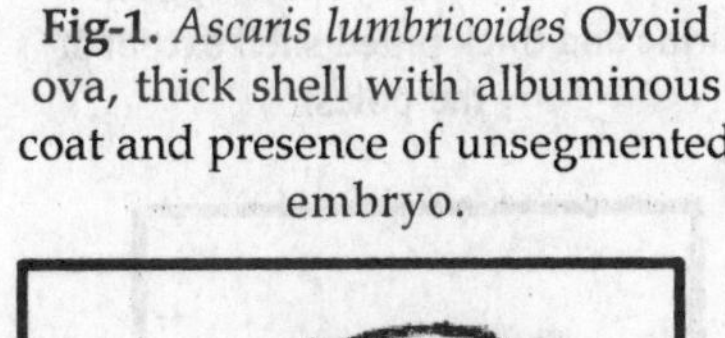

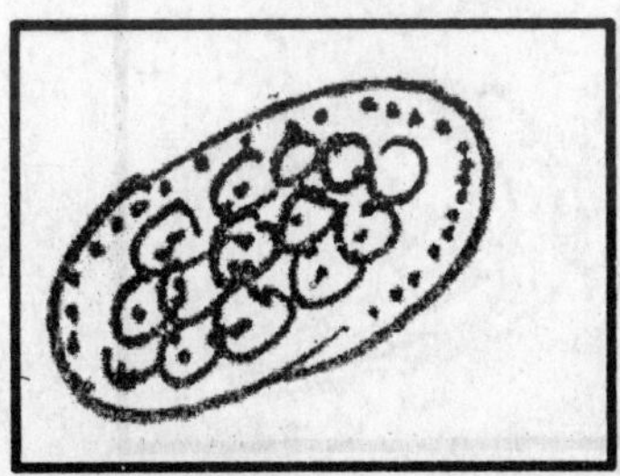

Fig-3 *Hyostrongylus rubidus* Ovoid eggs, thin smooth shell and presence of segmented embryo.

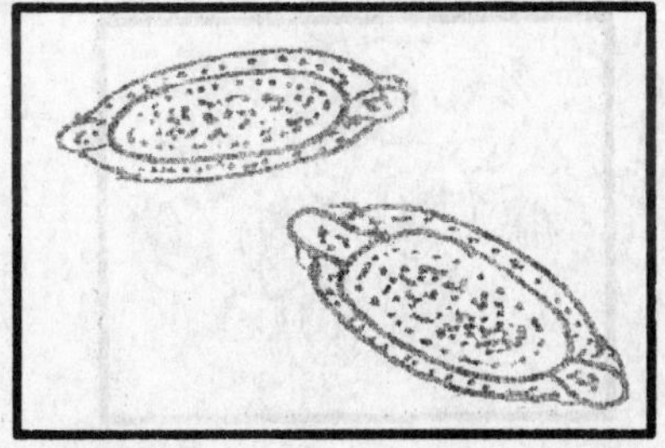

Fig-4 *Trichuris trichura* Barrel shaped eggs, brown shell with transparent plug at either end.

Fig-5 *Metastrongylus apri* Ovoid egg, thin smooth shell and presence of larvae.

Dogs

Fig-1 *Taenia hydatigena* Embryonated or unembryonated oval eggs or segments of parasites in faeces and oncosphere are with 6 hooks called hexacanth embryo.

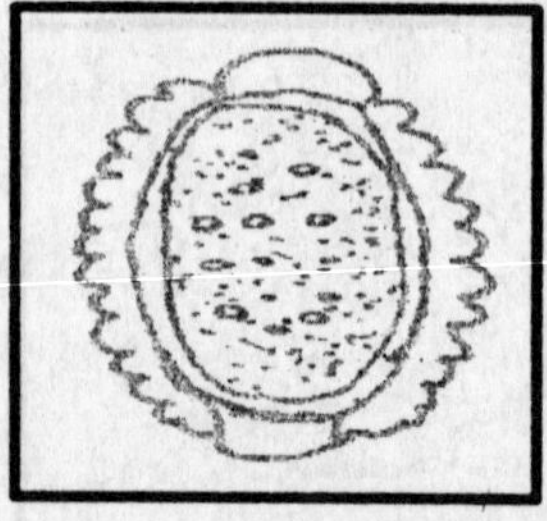

Fig-2 *Dioctophyma renale* Barrel shaped brownish yellow eggs in urine and thick pitted shell except at the poles.

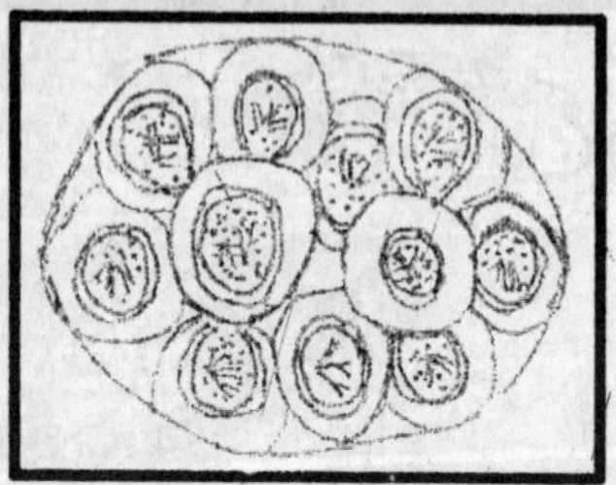

Fig-3 *Dipylidium caninum* Presence of thick eggs capsule with operculum containing upto 30 eggs and eggs released after its rupture.

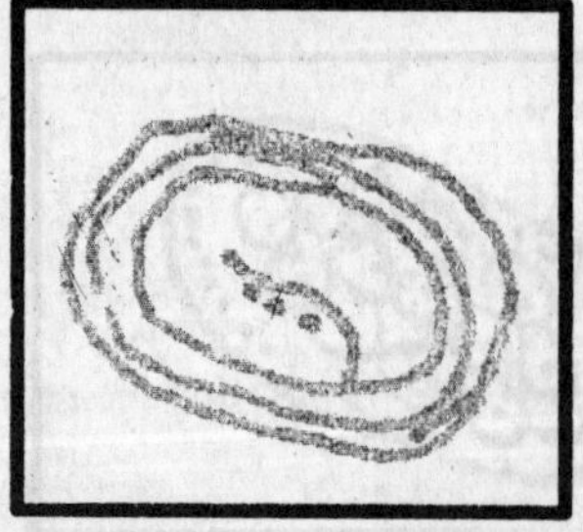

Fig-4. *Spirocerca lupi* embryonated eggs in the faeces. Nodules containing the worms in the oesophageal and aortic wall.

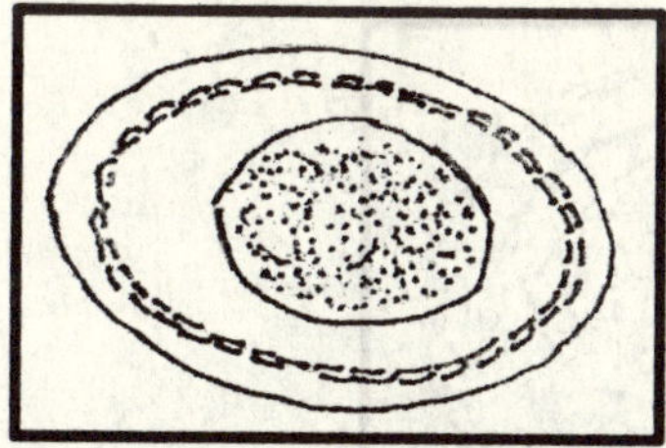

Fig-5 *Toxascaris leonina.* Smooth shell and ovoid eggs.

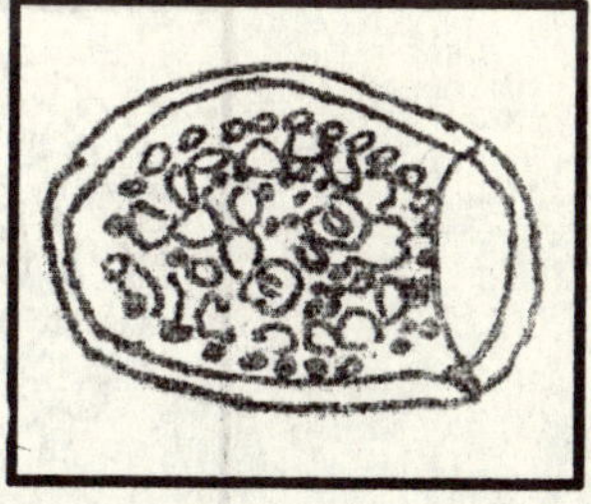

Fig-6. *Diphyllobothrium latum* Operculated ovoid eggs with rounded ends and golden brown colour

Fig-7 *Echinococcus granulosus* Features smilar to those of *Taenia spp.* Ova and segments noticed in faeces

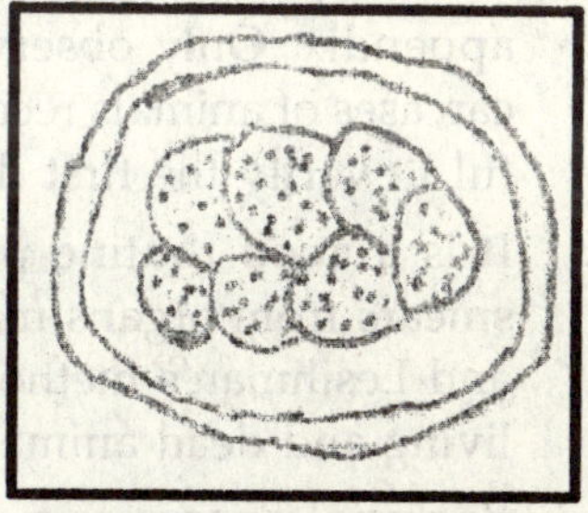

Fig-8 *Ancylostoma caninum,* A.braziliense (8a) and *Necator americanus* (8b) oval eggs, thin shell and presence of segmenting embryo

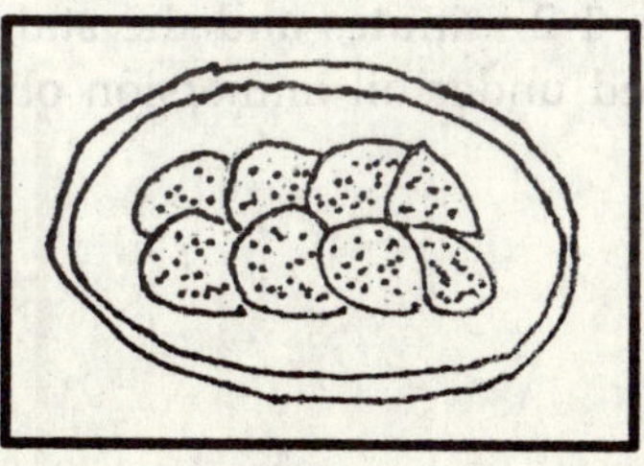

Fig-8 A

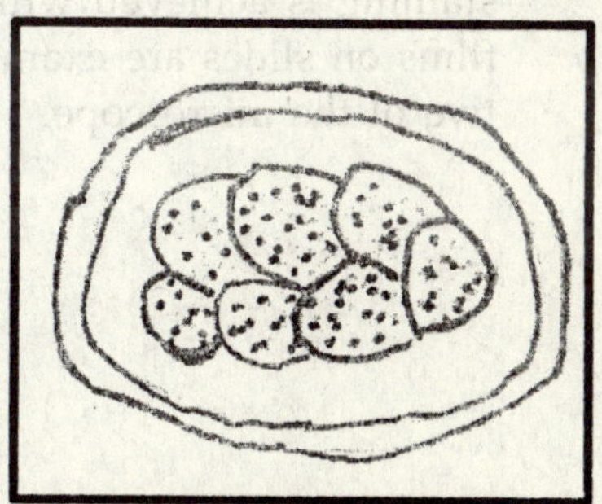

Fig-8 b.

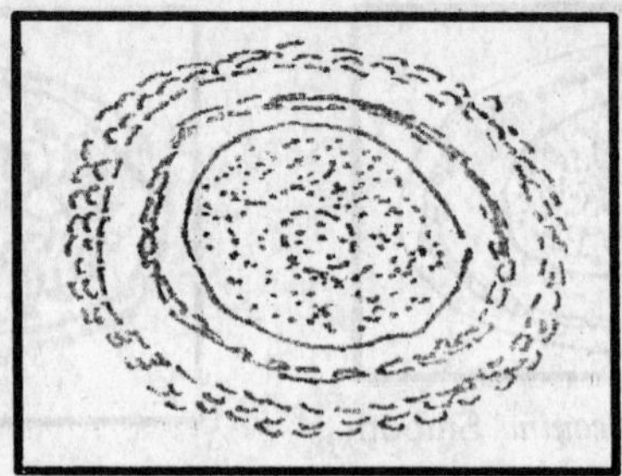

Fig-9 *Toxocara canis* (an ascarid) Round unsegmented and subglobular eggs, thick shell covering and dense embryonic mass.

1. For autopsy techniques in different species of animals, see appendix. Only observations of lesions in the organs of carcases of animals recorded by the first autopsist are helpful to verify the first diagnosis.
2. It is a good routine practice to stain the blood films or smears from organs made on clean slides by both Gram's and Lesihman's methods to detect bacteria or protozoa in living and dead animals etc.
3. Bacterial spores e.g. spores of ***Bacillus anthracis*** and ***Clostridium chauvoei*** are stained by Ziehl-Neelsen's method. The stained smears on the slides are decolourised with 5% acetic until the smears reveal light pink color. The counter stain is 1% methylene blue solution. The counter staining is achieved within 1-2 minutes and the stained films on slides are examined under oil immersion objective of the microscope.

Appendix

Postmortem Examination

It consists of systematic anatomical examination (i.e., first with the naked eye and, then, with the aid of a microscope) Bacteriological, parasitological and chemical examinations etc. are carried out to find out the definitive diagnosis as the cause of death. It gives an insight into the nature, extent and evolution of the disease in an organism.Obduction, cosmetic postmortem, complete and incomplete postmortems are some of the kinds of postmortem examinations which are carried out according to the object or purpose of the investigation. For example, an examination of only brain may serve the purpose as done usually in cases of rabies in animals (canines). Preautopsy data (e.g. age, sex etc.) enables the autopsist to adopt a more logical approach in discussing the priorities of different pathoanatomical diagnoses and the art of writing the epicrisis guides the autopsist to the definite cause of death after looking into details of merits of several primary, secondary and many other co-existing lesions in the dead animals. Postmortem protocol should be thoroughly prepared without missing to note any lesions and pathoanatomical diagnoses are made for different systems of the body in view of the lesions therein. The most important pathoanatomical diagnosis is given the first priority in declaring the cause of death. Deaths in animals may be either due to biological or aetiological factor. A rupture of heart, liver and thrombosis in cerebral artery may be responsible for biological death or somatic death of the animal. But aetiological diagnosis refers to a definite aetiological factor for brining about an end to life of an animal. For example, *Mycobacterium tuberculosis, Bacillus anthracis and Clostridium chauvoei* are considered as aetiological causes of death. The fact that an aetiological factor ultimately produces fatal biological or somatic changes in the body refers to another aspect of discussion.

Obduction refers to a medicolegal postmortem examination. The findings are used in the court for deciding the cause of death of accident or duration of some pathologic condition.

Cosmetic postmortem is a postmortem, which is done with least disfigurement to the cadaver. The parts are replaced in the body, the cuttings are sewed together and the body is washed and made to appear as nearly intact as possible.

Complete postmortem refers to an autopsy in which ever organ or part is carefully examined to make possible a correct diagnosis.

Incomplete postmortem refers to an autopsy in which a part of body (brain or limb etc.) is removed and examined for diagnosis of the disease. This is a sort of postmortem done to corroborate a postitive laboratory diagnosis.

Authority should be in a written form to do postmortem examination from the police officer, or the farm management or the owner of the dead animal to avoid unnecessary complications.

Clinical history (anamnesis) must be secured before autopsy of the dead body. Details of clinical diagnosis, treatment or other concerning facts are very much needed to arrive at a correct diagnosis postmortem.

Destruction

The dead animals after autopsy must be either burnt or suitably buried between layers of lime sufficiently deep in ditches dug for such purposes.

Care of the instruments

These must be sharp and properly sterilized to protect the autopsist during postmortem examination. The autopsists must be properly dressed i. e. there should be utilization of aprons, white coats, gumboots and gloves during autopsy.

Recoding the lesions

Postmortem findings must be recorded immediately after

autopsy in the postmortem hall by another person. There is usually a need of three persons for postmortem work.

A uniform and careful method should be followed for opening the dead body and examining the organs of the cadaver. To maintain the anatomical continuity of the organs of a system during postmortem examination is very helpful to make conclusions about the cause of death. This helps in an easy detection of an obstruction (e.g. stone or calculus) in the tubal structures (say. bile duct, urethera and parts of the gut etc.)

Postmortem protocol refers to a detailed written description of the postmortem findings. The protocol includes all information in history of the case or in relation to preautopsy data (e.g. bile duct, urethera parts of the gut etc.), findings of external and internal examination along with pathoanatomical diagnosis and epicrisi. Postmortem report refers to a brief report of the protocol and usually consists of pathoanatomical diagnoses as causes of death. A vetromedical report must be brief to prevent confusion in the court.

Pathoanatomical diagnosis refers to specific diagnostic terms in the listed form after external and internal examinations of different organs of various systems in the dead body. The terms like pneumonia, cirrhosis and nephritis are pathoanatomical diagnoses inferred from observations of the changes like degeneration, hemorrhages or inflammations in various organs.

Description of the lesions

In describing a lesion in a correct manner, descriptive terms such as colour, size, shape, consistency, odour, appearance of the cut surface and relationship of lesions to the surrounding structures are cautiously taken into consideration. A regular practice of this kind makes one a keen observer. Pathology is an observational science solidly helped by a through postmortem work, Preautopsy data refers to certain informations like time, place, position of the cadaver, weather, description of the agonal state, position of cadaver after death, owner, species, breed, sex, size and weight etc. It is very important to know the position of the animal before death and also after its death for correct interpretations of conditions like hypostasis or lividity etc.

Hypostasis or cadaveric lividity (livor mortis) refers to bluish red spots on the side upon which the animal has been lying down before death. This condition arises from an excessive accumulation of blood as seen veins of the most dependent parts of the body due to gravitation.

Epicrisis refers to the personal written opinion or idea of the autopsist about the entire case. The following facts are included in the epicrisis:

1. Pathogenesis of the disease
2. Cause of death.
3. Opinion about the primary lesions regarding its cause or development.
4. Statement concerning the importance of certain possibly co-existing lesions.

Rigomortis

It begins soon after death of the animal (say, after 4 to 24 hours). In some cases, it lasts usually for 24 hours. Sometimes, it may continue for 48 hours or longer and disappears with the onset of decomposition or putrefaction. It appears first in the eyelids, then, the masseter muscles and rest of the body and limbs and also disappears in the same manner. (i.e. disappearance seen first in the eye lids) It causes rise in temperature up to 3^0 F in the dead body of animals and reaches at maximum level after 20 to 24 hours and, then, it declines quickly.

It is characterised in dead animals by hardening and contraction of all the voluntary and involuntary muscles. Hardening of the muscles arises from coagulation of the myosin of muscles by **lactic acid** produced from muscle glycogen due to lack of oxygen. Rigomortis is related to break down of the muscle enzyme adenosine triphosphate (ATP) and the energy so released is used for the muscular contraction during rigormortis. It disappears due to the softening of the coagulated myosin by autolytic enzymes. An irreversible state of rigor motis reaches in the dead body of organisms, say, man when the level of ATP retained to 15% and lactic acid reaches a level 90.3 percent.

Rigomortis which takes place in heart earlier than its development in skeletal musculature is powerful enough in the left ventricle to express the blood from it. Some clotted blood occurs in normal heart but its presence **in the left ventricle** indicates incomplete rigor because of myocardial degeneration Unclotted blood in the left ventricle indicates hypoxia or disappearance of rigormortis from the left ventricle. Currant jelly clot or chicken fat clot in heart arises from the processes of sedimentation and blood coagulation. Red cells are present in the currant jelly clot but absent in the chicken fat clot.

Algormortis (L = coldness + of death) refers to a gradual drop in body temperature as compared to that of the surrounding atmosphere in dead animals.

In vetrolegal case, the veterinarians are confronted with the problems of differentiating antemortem wounds from postmortem wounds. Actually, there are no postmortem wounds or injuries, Wounds can only be produced during life of an animal.

Wounds or injuries like changes are made into the dead body of the animals with bad or criminal motives to mislead the autopsists. Organs also tear or rupture due to putrefaction or gas formation in the dead animals. In the dead animals, the main differences between antemortem and postmortem wounds like changes are given below:

Antemortem wounds	Postmortem wounds like changes
Inflammatory and reparative processes are seen in the body	Absence of inflammatory or reparative processes
Presence of gaping wounds due to stretched condition of the skin	No gaping of the edges is seen
Presence of the site of the wound	No clotting of the blood at the wound like sites
Spouting (i.e.) flow of blood with great force from arteries	No spouting of blood

Sometimes, the necessisty for deciding the age of injuries arises and the following facts help the veterinarians in arriving at a conclusion:-

I. Inflammatory change around the site of injury and changes in color indicate the occurrence of injury before death or probably 24 hours before death.

II. Hemorrhages and arterial spouting do not occur at the so called injured site after death of the animal.

III. Coagulation of blood occurs during life or within 10 minutes after death

IV. Edges of the wound may be averted or retracted and these changes occur during life or not within more than three hours after death Other useful factors in determination of the edges of the injuries are as follows:

(a) Healing of superficial cut in 10 to 24 hours

(b) Presence of inflammatory swelling within 36 Hours

(c) Union of the edges of the incised wound in 48 Hours

(d) Presence of complete healing in 4 to 7 days, leaving behind a tender scar.

(e) Growth of the granulation tissue to fill in the gaps of the wound in about 7 days.

Direct causes of death from wounds are haemorrhage, injury to vital organs like heart, lungs, brain, etc, and shock. A blow on the heart or abdominal region can cause shock in an organism and necrosis, septicemia, pyaemia, inflammation of internal organs, diseases, (e.g. tetanus,) are indirect causes of death. Cessation of respiration and circulation, cooling of the body, primary flaccidity, rigomortis, secondary flaccidity, putrefaction and mummification are the main signs of death in the animals. After a few hours following death, the body temperature falls and becomes equal to that of the atmospheric temperature. (i. e. cooling of body) Primary flaccidity (relaxation of the muscles) occurs immediately after death and lasts a few hours. Secondary flaccidity (relaxation of the muscles) becomes apparent with the onset of putrefaction and decomposition. The putrefaction is very rapid at 100^0 F and air and moisture promote putrefaction. The signs of putrefaction are as follows:

1. Abdomen distended with gas
2. Blood stained fluid from the mouth and nostrils
3. Liquefaction of eye balls
4. Presence of obnoxious odour or smell
5. Bursting of the abdomen and thorax with protrusion of stomach and intestine through it or eversion of the rectum through the anus.
6. Semifluid consistency of the tissues

A putrefied body is not suitable for postmortem examination but putrefied tissues can be used to perform some tests to ascertain the cause of death as done in the suspected cases of anthrax e.g. Ascolis test in the cases of anthrax.

A veterinarian is frequently requested to give wound certificate in medico or vetrolegal cases. The following form can be used for writing a wound certificate:

No. Date

This is certify that at the request of

(1) ....

I have this day examined.

(2) ....

Having the following identification marks ....

....

(3) ....

....

the said animal has got the following injuries on its body

(4) ....

....

I am of opinion ....

Place: Signature

Qualification

Designation

Autopsy techniques of different animals and birds

The postmortem techniques as followed by veterinarians in Sweden are very satisfactory, systematic and can lead to valuable findings in animal disease investigation. These methods definitely merit the acceptance and adoption in the veterinary Colleges in India. A mere observation of the lesions in the organs of body exposed by untrained persons is not a healthy practice.

The techniques to be adopted for postmortem examination in animals like cattle, horse, sheep, pigs and dogs etc. are given below:

Autopsy technique in large

ruminates (Rubàrth, 1964)

The technique in vogue in Sweden is as follows:

1. Scissors, knife, scalpel, forceps, sterile syringes, needles, vials, Petri dischaes, bone, tongs (shears), wax pencil and glass slides etc.
2. Large autopsy table, small autoclave, instrument boiler, metal trays, respiratory mask, apron hand gloves and gum-boot etc.

The body is supported on its back and inclined towards its left side. Evisceration of abdominal and thoracic cavities is performed from the left side.

The hind legs are abducted by cutting through the medical thigh muscles, opening the hip joints and cutting through the teres and accessory ligaments.

The udder is removed from females. In the case of males, the penis and prepuce are drawn backward to avoid being damaged when the abdominal wall is incised. This is done in the following manner. First, the parietal tunica vaginalis is incised and opened as far as the external inguinal ring and then the penis and prepuce are dissected free as far back as the ischial arch. The abdomen is then opened by an incision running along the line alba from the xiphoid cartilage backwards to the pelvis (do not extend the incision too far forwards otherwise the disphragm will be damaged).The abdominal wall is reflected by incising along the costal arch and, in females, along the anterior border of the pelvis. In males, the abdominal wall is incised from the midline towards the inguinal canal so that spermatic cord and the testicles can be freed.

After the abdomen has been opened and the abdominal walls reflected, the omentum is freed along its insertion to the lateral grooves of the rumen and from the duodenum, but the spleen is left attached to the rumen.

The fore-stomachs and abomasums are removed. The duode num is sectioned between two ligatures at the pylorus. Then, the rumen is freed from its attachments to the dorsal abdominal wall while the assistant pulls it over to the left side of the body. The esophagus is then cut through (be careful of the diaphragm), the forestomachs and abomasums are freed from their remaining attachment, and removed to the left side of the body.

The intestinal tract except for the duodenum is removed. After ligating the duodenum at the junction between its second and third parts, the duodenum and the pancreas are exposed and freed as much as is possible at this stage. The rectum is then freed. Ligated, and divided. The intestines are removed in one piece by cutting the mesentery along its attachment from the rectum and as far forward as the root of the mesentery. The duodenum and pancreas are allowed to remain in the abdomen for subsequent removal together with the liver. When the mesenteric root is cut through, the intestinal tract can be lifted out of the abdomen.

The urogenital organs, adrenals and rectum are removed. The floor of the pelvis is removed by sawing from the anterior and posterior borders through the obturator foramen on each side (do not saw too far literally).

The penis has meanwhile been freed from the ischial arch. The kidneys and adrenals are freed together and by carefully pulling them backwards as the urinary bladder. In females, the broad ligament is incised along its insertion on the abdominal wall so that the ovaries, Fallopian tubes, and uterus accompany the kidneys and ureters.All these organs are gathered into the left hand and then are drawn backwards at the same time as the attachments to the pelvic walls are cut through by cautious use of the knife. An incision around the anus completely frees these organs.

The abdominal aorta is opened in situ.

The diaphragm is cut through along its insertion beginning at the xiphoid cartilage (avoid incising the pericardial sac).Any pleural contents are collected and the amount is measured. The appearance of the pleura is inspected.

The pericardial sac is freed from the sternum by inserting the edge of the hand anterior to the pericardial sac and drawing the hand backwards. Keeping it as close to the sternum as far as possible. If any fluid is present in the pericardial sac it should be collected and the amount is measured.

The structures of the oral cavity and neck are then removed. An incision is made on each side of the tongue along the borders of the mandibles as far forward as the symphysis. The tongue is then pushed upwards between the mandibles, grasped and drawn backwards at the same time as the first incisions are extended backwards. The soft palate is cut free by incisions from each side which run forwards and medially to meet in the midline. The hyoid bones are divided at the joint between the main branch and the thyroid branch. The knife is placed at this junction with the sharp side upwards so that the great branch is lateral to the knife and the thyroid branch medial to it. If the knife has been correctly placed, a single sharp jerk will separate the bones. The pharyngeal structures are then dissected free laterally and dorsally. The trachea and oesophagus are freed to half way down the neck and then cut through.

Inserting two fingers through a short transverse incision

between two of the cartilage rings close to the free stump gives a good grip on the trachea which, together with the oesophagus, is then drawn backwards and freed as far as the thoracic aperture, After opening the aperture, the trachea and oesophagus are pushed into the thoracic cavity, The left hand is then inserted into the thoracic cavity, from the abdomen and the trachea is again picked up through the transverse incision. The trachea is then drawn backwards and the thoracic organs freed from the dorsal thoracic wall by cutting through the mediastinum.

The thoracic aorta, is removed together with the other thoracic organs.

The head with the salivary glands attached, is removed by cutting through the atlanto-occipital joint .

The body lymph modes are then examined. On the fore limbs (after these have been cut down), the prescapular, and auxiliary, are examined, on the hind limbs, the popliteal lymph nodes as well as the deep inguinal and the internal iliac lymph nodes should be checked for lesions.

The joints are opened from the medial surface beginning distally on the limbs.

The skeletal musculature is inspected by means of several incisions not forgetting the back and neck muscles.

The spinal cord is exposed by removing the dorsal arches of the vertebrae if examination is warranted.

AUTOPSY OF HORSE

The autopsy technique is much the same as in the cattle and buffalo except for the abdominal organs. The body is placed on its back and inclined toward its right.

Evisceration of the abdominal and thoracic cavities should be performed from the right side.

Inspect the contents of the abdominal cavity. The position of the organs is inspected, particularly the position of the left ventral and dorsal colon and the pelvic flexure. The diaphragm should be normally arched and tense, if it is not, ascertain the cause.

Winslow's foramen is located by following the visceral surface of the caudate lobe of the liver mediocranially. The foramen is a narrow slit, which will admit two fingers.

The great colon is lifted out of the abdomen to the right side of the body.

The small intestine is freed along its mesenteric attachment beginning with the duodenum at the site of the duodenocolic ligament. This ligament connects the third part of the duodenum and large intestines at the termination of the right dorsal great colon and the beginning of the small colon. The small intestine is divided between two ligatures and then is freed along the mesenteric attachment as far back as the ileum where it is divided again between two ligatures. During removal of the small intestine, the assistant should stand on the left side of the body and help by drawing out the intestine and tensing the mesentery.

The rectum, small and great colon, and caceum are removed together. The rectum is divided in the pelvis and the intestines are freed along their mesenteric attachment as far forward as the right dorsal great colon, In this region, the pancreas lies between the intestine and the sublumbar musculature and part of it lies under the intestinal serosa, Since the pancreas is to be left in the abdominal cavity (it is removed later together with the liver and duodenum) It is necessary to free the great colon from the pancreas and the dorsal abdominal wall by blunt dissection or by careful use of the knife in places. At this point only, the branches of the mesenteric blood vessels connect the great and small colon, caecum, and rectum to the body. These blood vessels are freed as far out as possible and then cut close to the intestinal wall. The large intestines are now completely freed and can be lifted out to right side of the body.

The spleen and omentum are removed.

The stomach is removed, the duodenum is divided between two ligatures immediately posterior to the pylorus. Then a short longitudinal incision is made through the serosa covering the portion of the oesophagus which extends into the abdominal cavity and the oesophagus is freed by blunt dissection, drawn out and cut off. The stomach is then freed of its ligaments and removed from the body.

The liver, duodenum and pancreas are removed together. The pancreas is first freed from the root of the mesentery and the dorsal abdominal wall and then the lateral and falciform ligaments of the liver are cut through. The incision through the falciform ligament is continued into the vena cava. The left wall of the vena cava is then slit by the insertion of the fingers while the liver is held ever towards the right side of the body. The liver is now attached to the body by only one wall of the vena cava and when this is cut through, the liver can be removed together with the pancreas and duodenum . Removing the liver in this manner avoids damage to the diaphragm.

The mesentery and the abdominal aorta are removed. The aorta is divided immediately behind the diaphragm and in front of the mesentery by drawing gently backwards, it can be freed from the dorsal abdominal wall. The aorta, with mesentery attached, is then removed by cutting the iliac branches as far distally as possible (about 10 cm)

AUTOPSY OF SMALL RUMINANTS

The technique is similar to that for larger ruminants except that the sternum is removed by cutting through the costochondral junctions. The neck and thoracic organs are then removed together, i.e. without cutting of the trachea and oesophagus anterior to the thoracic aperture

AUTOPSY OF CARNIVORES

On the whole, the autopsy technique for these species follows the description described above but with the following differences.

Examination and removal of the abdominal organs.

The incision along the linea alba should not extend farther forwards than the xiphoid cartilage, otherwise the diaphragm can easily be damaged . The omentum with the spleen is lifted up and folded over the thoracic wall so that the position of the abdominal organs can be inspected.

The intestinal tract is removed , Divide the rectum in the pelvis, after ligation, if necessary, and then work forwards by cutting along the mesenteric attachments, Continue until the tip of the duodenal branch of the pancreas is reached, this represents the junction between the duodenum and jejunum, Ligate the small intestine at this point and divide it.

The omentum and spleen are removed along the insertion of the omentum, taking care to avoid the splenic branch of pancreas.

The liver, stomach, pancreas and duodenum are removed together.

Begin by freeing the pancreas from the root of the mesentery and the dorsal abdominal wall. The ligaments of the liver are then freed, to avoid damaging the diaphragm follow the contour of the liver very closely. The stomach is, then, cut free through the cardiac region, the incision should not be made too far cranially otherwise the pleural cavities will be opened. Then cut through the vena cava to free the liver completely. After breaking down or cutting through any remaining ligaments, the organs can be removed in one piece.

The autopsy is then continued as for the cattle with the exception that the sternum is removed by cutting through the costochondral junctions before taking out the neck and thoracic organs in one piece.

AUTOPSY OF PIGS

The technique is similar to that applied to carnivores except for removal of the intestinal tract.

To remove the intestinal tract begin at the tip of the colonic spiral and free the entire intestinal tract along the mesenteric attachments orally and aborally.

In large pigs, the stomach is removed separately, in smaller animals, the stomach is removed together with the liver, duodenum and pancreas.

AUTOPSY OF BIRDS

1. The legs are abducted by cutting or breaking opens the hip joints. The abdomen is opened and the sternum is freed by lateral incisions through the ribs. The crop is then freed be blunt dissection from its attachment along the thoracic aperture to avoid being damaged when the heavy anterior osseous attachments of the sternum (the coracoids and clavicle) are clipped through with bone tongs on each side to join the lateral incisions through the ribs.

2. The abdomen is opened and the sternum is freed by lateral incision through the ribs. The crops are, then, freed by blunt dissection from its attachment along the thoracic aperture to avoid being damaged when the heavy anterior osseous attachments of the sternum (the coracoids and clavicle) are clipped through with bone tongs on each side to join the lateral incision through the ribs.

3. The spleen is removed separately.

4. The stomach and intestinal canal are removed in one piece after cutting through the oesophagus just anterior to be proventriculus.

5. In sexually mature females, the ovary is removed at its base and the oviduct is first extended by cutting through its dorsal and ventral mesenteric attachments and then removed by cutting through the cloaca.

6. The pericardial sac is incised and the chambers of the heart is opened in situ by incising the wall of the right ventricle near the apex and continuing the incision interiorly up through the pulmonary artery and laterally up through the right atrium. The procedure is repeated for the left ventricle extending and

incision at the apex up through the aorta and up through the left atrium. The heart is then removed by cutting through its base.

7. The lungs are freed by blunt dissection from the thoracic walls, cutting through the dorsal attachment (dorsal to the thoracic oesophagus and aorta), and then removed by cutting through the trachea immediately anterior to the syrinx.

8. The upper beak is cut transversely at its base to expose the nasal cavities, and then the mouth is opened by cutting through one corner (the right is most convenient) and the incision continued through the pharynx and down the oesophagus to open the crop, The trachea is then opened along its whole length.

9. The brachial plexus and the sciatic nerve are exposed on both sides.

10. The major joints are opened.

Precautions

When a case of a dead animal is presented for postmortem examination, there are certain extremely important precautions to be remembered and followed during autopsy work. Postmortem work done in a haphazard manner will lead one to no conclusion with only waste of the material and time.

The precautions are as follows:

1. Consideration of a postmortem case should be made as a serious bit of research work and anamnesis (history) of the case should be read carefully. If necessary, more information about the dead animal can be obtained from either the attending veterinarian or owner of the animal.

2. Special attention is given to the organs in the dead body of the animals which have shown clinical signs of abnormality.

3. Lesions in the bodies should be noted as encountered during postmortem work. Postmortem findings should never be written from one's memory after the work is over or even after lapse of a day or two. There is always a risk of forgetting

very valuable findings after passage of sometime from the moment of postmortem examination.

4. Materials for bacteriological, parasitological and chemical examination etc. should be removed and properly preserved for further laboratory investigation.

5. The skin of the dead animal should be carefully examined on both sides of the body and distinguishing marks or any abnormalities e. g. discharges from natural orifices should be recorded.

6. The dead body should be examined to note the state of nutrition, inanition and anaemia etc.

7. The autopsist must obtain the authority to do postmortem examination from police, attending doctor or the owner of the animal and he should be properly dressed for this kind of work. Gum-boots, white coat, apron and rubber gloves should be used invariably as a routine measure before commencement of the autopsy work.

8. Nutritional state (obesity vs. emaciation) and wasting conditions affect the relative weight of the organs in dead animals.

The formula for calculating relative weight in dead animals is as follows:

$$\text{Relative weight} = \frac{\text{Weight of organ}}{\text{Weight of the body}} \times 100$$

9. Descriptive terms liks size,shape,weight, colors, dour consistency, incision contents appearance of cut surface, and relationship of the lesions to the surroundings are to be used in depicting the changes or lesions in the organs of the bodies in a careful and objective manner as they are exposed or encountered while examining the dead bodies postmortem.

10. A first aid kit should be readily available in the postmortem room to treat minor cuts or pricks incurred during autopsy.

11. Gastrointestinal tract should be left to the last for the sake of examination during autopsy in order to avoid contamination or soiling the organs or materials needed for diagnostic purpose.

12. For collecting the fluid or blood for cultural examination heart wall should be first seared with red hot spatula and the fluid is drawn into sterilized capillary pipettes.

13. During the postmortem examination, films and cultures are made from the different fluids and tissues. The surface of the organ should be seared with red hot spatula in order to scrape parenchyma with sterilized scalpel. The material is picked up with sterilized platinum loop in the hot flame for bacteriological examination using suitable media, Cultures can be made from bone marrow in the animals which have been dead for a long time. The bodies of such animals are usually in very advanced stage of decomposition.

14. The various inoculation routes for administering the extract of lesions in the animals are subcutaneous, intracerebral, intravenous, intraperitoneal, intradermal, intratesticular and intraocular etc. The inoculation site is clipped or shaved and cleared with cotton soaked in ether or spirit. For collecting blood, the needle is inserted in to the vein in the direction of blood flow.

15. All the contaminated instruments and discarded cultures etc. should be placed in big jars containing disinfectant solutions like 5% carbolic acid or 1 % lysol solution in water.

16. A long dead animal is useless for any bacteriological examination.

17. Blood films for microscopical examination are made thin ones as far as possible. Dry the blood films immediately by waving them in the air. Prepare thick blood films in cases of anthrax by simply smearing a drop of blood on a clean slide with the help of a platinum loop. Intercellular parasites (say. *Babesia sp.*) and leucocytes are found easily in large numbers at the edges or termination of the blood films.

18. Sputum for detecting tuberculosis organisms may be

collected from the wall etc. in front of suspected TB cases The animal can be made to cough on a piece of a large sheet placed over the wall. Thick or turbid mucus from TB patients is quite satisfactory.

19. Nematodes should be collected in a test tube containing normal saline solution(0.85% Sodium chloride in distilled water) and the tube is shaken well to clean the parasites. Hot 70% alcohol containing 3 to 5 % glycerine is poured over the worms to kill and fix them. The worms get cleared after sometime in this preservative.

20. Materials from animal showing symptoms or representative lesions of bacterial, viral or parasitic diseases etc. are collected in the conditions as given below :

I. Acute stage of the disease

II. Immediately after death of the animal

III. After sacrificing the diseased animal

21. Disposal of the dead bodies of the animals is done by burial of the dead body along with dungs, bedding blood stained soil or discharges in order to prevent the spread of outbreak of diseases (e.g. anthrax). Burning of the dead bodies may be disadvangeous owing to the high cost of the fuel. But one should be extremely careful in burying or disposing of the cadavers. At first, a trench of the adequate size with a depth of at least 6 feet is prepared by digging the earth. A foot of lime is layered over the trench and the carcass is laid over it. The whole dead body must have a covering of a foot of lime from all sides (i.e. top, sites and base etc.). If possible, one can burn the cadaver using an incinerator .Spores of anthrax are known to be spread by earthworms from trenches because of improper burial of dead bodies of anthrax cases.

The lack of oxygen in the dead body of anthrax causes disintegration of anthrax bacilli in the body in about few days depending upon the climatic conditions.

B. Procedures, stains and reagents commonly used in the laboratories

These are as follows:

1. Leishman's Stain
2. Giemsa's Stain
3. 1% Crystal violet solution
4. Ziehl' Nielsen's carbol solution
5. 1% methylene blue solution
6. Gram's iodine
7. 2-3 % Acid alcohol
8. 95% Absolute alcohol

Films or smears of blood and exudates etc. are made on clean glass slides to detect bacteria and protozoa etc. Thin films are usually prepared but thick films are made in some disease (for example, anthrax). Details of the stains and procedures followed are given below:

(A) Leishman's stain

It is prepared by dissolving 0.15g Leishman's powder in 100cc pure acetone free methyl alcohol. It is quite good for differential count as well as for staining bacteria and protozoa in films and sections.

Buffer $Na_2 HP_4 12H_2O$ (sodium phosphate) 35.61 g to litre

$NaH_2 Po_4 2H_2O$ (sodium dihydrogen phosphate) 27.61 g to litre

Autoclave it at 15 Ibs (pressure) per sq. inch for 30 minutes. And keep it in cold stores.

Procedure

1. Pour the Leishman's stain on the unfixed smear or blood films and leave it to act for 1 minute.

2. Add double the volume of freshly prepared distilled water or buffer solution (i.e. 1 part of the stain and 2 parts of distilled

water) at neutral pH on to the smears on this slides and mix the stain and buffer with a Pasteur pipette by sucking it up in the pipette and expelling it out on the slide.

3. Allow the diluted stain to act 5 to 10 minutes depending upon the quality of the Leishman's stain.

4. Examine under the oil immersion objective after taking one or two drops of cedar wood oil or even liquid paraffin. Intercorpuscular parasites (e.g *Babesia* spp.) are usually seen at the margins and termination of the films in mammals, Leucocytozoon in peripheral blood films of chickens is stained with brilliant cresy1 blue. Plasmodium is stained in the red cells by any Romanosky stain. The initial stage of the merozoites occurs in the ring forms in the red cell and a vacuole is noticed in the parasites. Haemoproteus infects red cells of pigeons. Trypanosomes also infect domestic and wild birds.

(B) Giemsa's stain

It is prepared by dissolving 3.8g of Giemsa stain in 250 ml pure methy1 alcohol (acetone free) by shaking for 15 minutes and then it is added to 250ml of glycerine. The mixture is shaken for 10 minutes. Filter the stain and discard the residue. It keeps very well and is good for bacteria and protozoa in the smears. It can be used for staining Pasteurella, Chlamydial inclusion bodies, haemophilus and spirochaetes.

1. Fix smear or blood film with methyl alcohol for 2 minutes.
2. Stain with dilute Giemsa's stain (2 drops of stock Giemsa's solution in 1 ml of buffer, pH6.8 or freshly prepared distilled water) for 30 minutes to one hour in view of the quality of the stain.
3. Wash with the same buffer solution or distilled water.
4. Dry in the air, add a drop of cedar wood oil on the smear and examine under the oil immersion objective.

(C) Gram's stain

Solutions required:

1. 1% Crystal violet solution
2. Gram' iodine or Lugol's iodine solution lodine 1g, potassium iodine 2 g and water 300 ml.
3. Dilute carbol fuchsin.

It is prepared by mixing 1 part Ziehl Neelsens stain with parts of distilled water.

4. 95% absolute alcohol

Procedure

i. Smears from tissues or exudates are made, dried in the air and fixed by heat.

ii. Stain smears for 1 minute with 1% crystal violet solution.

iii. Wash the smears in tap water for about 2 seconds and add iodine solution (a mordant) on the smears and allow it to act for 1 to 2 minutes.

iv. Wash in tap water and decolorize the smears with 95% absolute alcohol for 30 seconds and wash again the smears in tap water.

v. Counter stain the smears with dilute carbol fuchsin for 10 seconds.

vi. Wash in tap water, dry and examine under oil immersion objective.

Note: - Gram Positive bacteria stain blue or violet but the gram negative organisms stain red or pink.

Ziehl-Neelsen's stain

Good for acid fast bacteria or spores in the films or sections.

Solutions required are:

1. Ziehl- Neelsens stain It is prepared by dissolving one gram basic fuchsin in 10 ml absolute alcohol and 100ml of an aqueous solution of carbolic acid(1 to 20) is then added to it. The dye is dissolved in the absolute alcohol and, then, the dye in alcohol is added to the phenol solution in the distilled water.

2. 3% acid alcohol
   Hydrochloric acid 3ml
   70% alcohol 97 ml
3. 1% aqueous solution of methylene blue

Procedure

(1) Stain the dried and fixed smear with Z.N stain. Smear is flooded with the stain and is heated until the steam rises and the hot stain is allowed to act for 2 minutes. Heat can be applied at intervals to keep it hot.

(2) Wash with water.

(3) Decolourise the smears, until the smears are faintly pink.

(4) Wash the smears in tap water and counter stain with 1% methylene blue for 10 to 20 seconds.

(5) Wash, dry and examine it under oil immersion objective after adding one or two drops of liquid paraffin on to the smears.

FAECAL EXAMINATION

(1) Direct method

One or two loopful of the faecal matter is mixed with water on a slide.

A loopful of the diluted sample is then covered with a cover slip and examined under low and high power objectives of the microscope. The method is very useful in diagnosing coccidiosis in the fowls. Duodenal or caeccal blood tinged scrapings can be used to detect coccidian. Suspended oocysts, merozoites and developmental stages of coccidian in the epithelial cells are noticed in direct wet mount preparations of intestinal mucosal scrapings of the infected birds.

(2) Concentration methods

(a) Sedimentation technique

The faecal sample is mixed with water and centrifuged. After discarding the supernatant, a loopful of the sediment is taken on a slide and covered with a cover slip to examine the sediment under low and high powers of the microscope.

(b) Floatation technique

(1) 2 to 3 gm of the faecal samples are liquefied in water and shieved to remove the coarse particle.

(2) The liquefied sample is poured into a centrifuge tube and centrifuged for 2 to 3 minutes at 1000 to 1500 rpm.

(3) The tube is taken out and the supernatant fluid is discarded and tube is filled with sugar solution. The sugar solution is prepared by dissolving 1 Ib of sugar in 12 ounces of water under heat. A few drops of phenol (1 % in water) is added to it to act like a preservative.

(4) The sugar solution and the sediment in a centrifuge tube is mixed by inverting the tube twice or thrice and the tube is again centrifuged for 3 to 5 minutes.

(5) A drop of the supernatant fluid is taken on a slide and covered with a cover glass. The preparation is then examined under low and high powers to detect ova or eggs of the parasites.

G. Examination of Ectoparasites and Fungi Etc.

(a) Mites (the causative factors of cutaneous infestation or mange) insects, ticks and mites etc are carriers of many pathogenic viruses, protozoa and spirochaetes in birds. Mites causing cutaneous lesions are noticed in the skin lesions or crushed nodules under a cover slip in one or two drops of acidulated water. Air sac mites (*Cytoleiclius spp.*) are detected in the air sacs and respiratory tract of the affected birds. Mange mites of the genus *Cnemidocoptes* are parasitic for only birds. Skin scrapings collected from the edge of astive lesions are boiled gently for 5 minutes in 5 % KOH solution and then centrifuged at 1500 r.p.m for about 5 to 10 minutes. The deposits are examined under low and high powers to detect the mites.

(b) Some ectoparasites are seen in the internal organs of birds at autopsy.

(c) Examination of fungi (mycotic agents)

Fungi (dermatophytes) are noticed in the scabs, crusts or epithelial cells in the cutaneous lesions. The crusts or epithelial cells in the lesions are placed on a slide in 10 to 40% solution or potassium hydroxide for microscopic examination under the low and high power objectives, *Trichophyton gallinae* (the causative agent of favus in chickens and turkeys are found in the epidermal cells of scrapings. Interwoven hyphae with arthrospores are found in the scabs or lesions of comb or scaly lesions on the different parts of the body of infected birds.

Some useful definitions about diseases are given below:

Epizootic refers to an attack of a rapidly spreading disease simultaneously in a large population of animals. An enzootic differs from it in that it is continually present in some areas and is not introduced from outside. Some times, an enzootic may turn into an epizootic (e.g.FMD). Incidence means the number of cases of a disease in an animal population.

Incidence rate (IR)

It denotes the number of new cases of a particular disease occurring in a given population at a stated period of time.

The morbidity may refer to an incidence or a prevalence rate.

C-SMR (Cause specific morbidity rate) is expressed as given below.

$$\text{C}-\text{MSR (IR)} = \frac{\text{New cases reported during a period}}{\text{Average population during that period}}$$

Prevalence rate denotes a static measure of total number of the affected individually in a population at a given time.

$$\text{Crude death rate(CDR)} = \frac{\text{Deaths in a year per thousand}}{\text{Population at mid year}}$$

Case fatality ratio

$$\text{(CF ratio)} = \frac{\text{Number of deaths in a particular case}}{\text{Number of case of the same cause}} \times 100$$

POSTMORTERM REPORTS ON ANIMAL DISEASES

Postmortem examination of the dead animals is a routine work for the field veterinarians. How to record the lesions seen in the dead bodies during autopsy was described earlier. The postmortem reports on 30 important diseases of domestic animals have been narrated to inculcate bases in the minds of students and veterinarians to correlate the symptoms and lesions of the diseases for the sake of preparing the autopsy reports on the dead animals, which succumbed to various ailments. There is a close correlation (barring a few cases) between antemortem symptoms and postmortem lesions of diseases caused by various factors. Significant lesions are not noticed in a few diseases marked by nervous symptoms (for example, tetanus and strychnine poisoning etc.) Calculation of the relative weights gives an assessment about the nutritional state (obesity vs. emaciation) of the dead animal.

Approximate relative weights of organs in health are the following:

Organs	Pigs	Horse	Cattle	Dogs
Heart	0.75	0.4	0.8	0.75-3
Lungs	0.85	0.7	1	--
Liver	1.00	1.25	2.5	1.5
Spleen	0.2	0.2	0.2	0.15
Kidneys	0.25	0.25	--	0.5

The diseases chosen for describing the postmortem reports are as follows:-

1. Anthrax
2. Haemorrhagic septicaemia
3. Black quarter
4. Surra
5. Rinderpest
6. Theileriasis
7. Babesiosis
8. Anaplasmosis

9. Tetanus
10. Brucellosis
11. Foot and mouth disease
12. Rabies
13. Haemorrhagic enteritis
14. Tuberculosis
15. Johne's disease
16. African horse sickness
17. Navel- ill
18. Coccidiosis
19. Colibacillosis (White scour)
20. Swine fever
21. Poisoning from ingestion of the organophosphorous compounds.
22. Snake bite
23. Secondary shock from an accident
24. Burn shock
25. Lightning injury
26. Drowning
27. Traumatic pericarditis.
28. Trichostrongyliasis
29. Canine distempher (C D)
30. Photosensitization

The postmortem reports on the aforesaid diseases are not meant for copying in any way by a veterinarian but can be useful in preparation of a complete postmortem report and for comparison of the lesions seen in the dead bodies in order to make a safe diagnosis. The field veterinarians may, however, give only presumptive postmortem diagnosis (i.e. a diagnosis required to be confirmed by laboratory findings) with despatch of the materials to a well equipped laboratory for reports of bacteriological, parasitological and other tests etc.

Postmortem No. 1	**Breed A cross-bred cow**
Species Gray	Age : About 4 -year-old cow
Owner:M.N. Ray	Weight: 600 kg
Residence	

By whom sent for examination and reasons, if any:

Veterinary Assistant Surgeon, Phulwari.

To ascertain the cause of death.

Date and hour of death	Autopsy findings:
6 a.m at 3.5.1990	1. Tarry (dark colored) unclotted blood at the cut ends of the vessels
Date and hour of Postmortem	2. Absence of rigor mortis
Examination	3. Enlargement of spleen (splenomegaly)
11 a.m on 3.5. 1990	4. Petchial heamorrhages in the intestinal mucosae and heart etc.

History:-

Deaths of cattle reported all on a sudden within 1-2 hours or a day from the onset of illness. Fever recorded in the patients up to 106^0 F. Respiratory distress and muscles tremors noticed in several animals. Blood tinged discharges were seen at the orifices like mouth, nostril, anus and vulva. Oedematous changes in the tongue and sternum in some cases. Only one dead case war carted to the department for autopsy.

Cadaverous changes – Bloody discharges at the nostrils, mouth, anus and vulva. Conjunctivae of the eyes were reddened.

External appearance on the cut removal of skin (describe Contusions and wounds etc.)	Unclotted tarry blood at ends of the blood vessels. The carcase was partly decomposed.

Mouth and pharynx : Pharyngeal mucosae congested.

Nasal cavities : Nasal mucosa congested and covered with blood tinged fluid.

Larynx and trachea : Moderate congestion and a few petechiae in the tracheal mucosa.

Pleural cavity and lungs: Modertely congested lungs.

Pericardium and heart (absolute wt.... relative weight .... %)

Subendocardial and subepicardial haemorrhages.

Peritoneum Haemorrhages on the peritoneum. No blood clots were found in the opened blood vessels.

Liver (Absolute wt 5 kg. Relative wt 0.83 %)

Reddened blood escaped from the cut surfaces.

Gall bladder – N.A.D (No abnormality detected)

Spleen (Absolute wt. 2.5 kg. Relative wt. 0.41%).

Severely enlarged and reddened dark.

Stomach and intestine : Mucosae of the stomach and intestine congested and showed haemorrhages.

Kidneys : Swollen and reddened (absolute wt 1.5 kg. relative wt 0.25 %)

Generative organs - N.A.D

Brain and spinal cord - Not opened

Lymph glands in general : Enlarged and congested.

Blood or tissue smears sent for microscopical examination:

1. Smears from the cut tips of the ear.
2. Smears from the spleen.
3. Smears from the heart blood.

Diagnosis on the basis of above examination: Anthrax

Remarks (State if viscera and tissue specimens etc. sent for chemical, bacteriological, histopathological and parasitological examination)

Microscopical examination of blood etc.

Results of chemical, bacteriological, parasitological, and histopathological examinations.

1. Gram positive rods indistinguishable from *Bacillus anthracis* were detected.

Place where postmortem examination was done ....

Dated:

Note: - Dead bodies of the suspected cases of anthrax are not advised to be opened under the field conditions. It is better to despatch the blood films from the blood escaping from the cut ear veins. Decomposed pieces of bones or organs may be despatched to the laboratory through a courier for Ascoli's test. Autopsy of anthrax may be undertaken in a postmortem laboratory with adequate arrangement for incineration of the carcases in order to prevent the spread of the disease to man and animals

Sd/-

A. Prasad

Signature of the autopsist

Designation:

Postmortem No.2

Species – Bovine	Breed : Murrah
Colour – Black	Sex : Female
Owner – A.N. Lal	Age : 5- year old buffalo
Residence , Araah	Weight : 800 kg.

By whom sent for examination and reasons. If any:

Touring Veterinary officer

To ascertain the cause of death.

Date and hour of death: 6.6.90 at 2 am.	Autopsy findings: 1. Oedematous swellings in the neck and sternal region.
Date and hour postmortem Examination 6.5. 90 at 9 am	2. Reddened and swollen lungs. 3. Subepicardial and subendocardial haemorrhages. 4. Haemorrhagic congestion in the abomosum and intestines.

HISTORY

Several deaths of young buffaloes with rise of temperature (up to 107^{0}F) and hot subcutaneous oedematous swellings were noted in sternal, neck and intermandibular regions. Such illness started with the onset of rain during the summer days and many affected animals died within 24 hours. Swollen tongues and oedema in the pharyngeal mucosa were noted in some affected animals. Constipation followed by haemorrhagic diarrhea was also noted in the sick animals. Only one dead body was received for autopsy.

Cadaverous change : Subcutaneous swellings in the neck sternum and intermandibular spaces. Conjunctivae of the eyes congested.

Rigor mortis – Present in the fore-limbs. External appearance on removal of skin (Describe contusions and wounds. etc.)

Reddened oedematous swellings in the throat and neck region marked by formation of pits on application of force. Muscles dry and sticky to the touch.

Mouth and pharynx :- The pharynx was reddened.

Nasal cavities – N.A. D.

Larynx and trachea – The mucosae of the air passages reddened and revealed haemorrhages.

Oesophagus - N.A.D.

Pleural cavity and lungs– Both the lungs were reddened and swollen. A few haemorrhages on the pleural surface.

Pericardium and heart (absolute wt .... relative wt .... %)

Subepicardial and subendocardial haemorrhages.

Peritoneum – Haemorrhages on the peritoneal serosa.

Liver (Absolute wt 6 kg. relative 0.75 %) – Swollen and reddened. Blood escaped from the cut surfaces.

Gall Bladder – Thick bile

Spleen (absolute wt1.5 kg. relative wt 0.18 %) – Reddened.

Stomach and intestine- Abomasal and intestinal mucosae, reddened, swollen and covered with blood tinged material.

Urinary organs –

Kidneys were reddened and swollen. (Absolute wt 2 kg. relative wt 0.25 %)

Generative organ – N.A. D.

Brain and spinal cord – Not opened

Lymph glands in general – Swollen and congested.

1. Smears from the heart.
2. Smears from the spleen.
3. Smears from the lymph nodes.

Diagnosis on the basis of -

above examination Hemorrhagic septicemia.

Remarks (state if viscera and	A. Tissue pieces of spleen and lungs in sterile Petridishes.
tissue specimens etc. sent for chemical, bacteriological, histopathological, histopathological and parasitological examinations)	

Microscopical examination of blood etc

Results of chemical,	1, 2 and 3, a few G- bipolar
bacteriological, parasitolo gocal and histopathological examination	organisms indistinguishable from

Pasteuralla multocida were detected.

Place where postmortem examination was done

Note: Pasteuralla organisms were isolated. Haemorrhagic tracheitis in the rabbits was produced following injection of the isolate.

Dated

Sd/- A. P Roy

Signature of the autopsist

Designation.

Postmortem No. 3

Species –Bovine

Colour – Red

Owner – S.B. Singh   Breed – A crossbred cattle

Residence – Patna   Sex – male

Age- 6- year-old bullock

Weight – 400 kg.

By whom sent for examination and reasons, if any:

Touring veterinary officer

To ascertain the cause of death

Date and hour of death Autopsy findings

0.6.91 at 2.00 p.m Date and hour of post mortem examination 10.6.91 at 3.00 p.m.	1. Dark red swellings in the thigh muscle infiltrated with blood and gas bubbles.
	2. Blood stained fluid at the nostrils 3. Rapid putrefactive change and abdomen distended with gas.

History:

Lameness with rise of temperature (upto 106^0 F) and rapid pulse (upto 120 per minute) were reported in young cattle under good nutritional state. Hot, oedematous, painful swellings with crepitating sounds were noticed in the thigh and shoulder, muscles of the sick animals. In some animals, such emphysematous swellings were cold and painless on palpation. Deaths were noted in such cases within 26 hours after the appearance of symptoms. Only one carcase was received for postmortem examination. Slides prepared with fluid obtained from swellings by needle puncture were also sent for examination.

Cadaverous changes- Swellings on the thigh and back quarter. Blood stained fluid at nostrils.

Rigor mortis – Present in the head and fore- limbs.

External appearance on removal of skin (describe contusions and wounds etc.)	Crepitating swellings on the upper part of the left thigh in the muscles. Blood stained fluids mixed with a few air bubbles escaped from the cut surfaces. Partly decomposed carcase, Presence of rancid odour in the opened swellings.

Mouth and pharynx – N.A.D

Nasal cavities – N.A.D

Larynx and trachea - N.A.D

Oesophagus – N.A.D

Pleural cavity and lungs- N.A.D

Pericardium and heart (absolute wt ....relative weight .... %) A few subendocardial petechiae.

Pertitoneum- Partly clotted blood in the left ventricle. About a Ib of serosanguinous fluid.

Liver (absolute wt............... relative wt .... %)

Partly decomposed and swollen

Gall bladder- N.A.D.

Spleen – N.A.D (absolute wt. .... Relative Wt .... %)

Stomach and intestine- Partly decomposed.

Urinary organs.

Kidneys (absolute wt........relative wt....................%)

Swollen and congested.

Generative organs – N.A.D.

Brain and spinal cord - Not opened

Lymph Glands in general – Prescapular lymph nodes swollen and reddened.

Blood or tissue smears sent

For microscopical exam. 1. Smears from the thigh swellings.

Diagnosis on the basis of above examination – Black quarter (B.Q.)

Remarks (state if viscera and tissue specimens etc. sent for chemical,	A. Tissue pieces from the muscles on the ice in the sterile Petridishes.

of the bacteriological, histopathological and parasitological examinations	B .Smears prepared from the fluid muscular swellings obtained by needle puncture.

Microscopical examination of blood etc.

1. Boat shaped sporulated organisms indistinguishable from *Clostridium chauvoei.* Subterminal spores also noted in the rods.

Results of chemical Bacteriological, parasitological and histopathological examinations	A. The Clostridial organisms were isolated and seen in the smears of livers of the guinea pigs injected with the isolated
	B. Gram positive boat shaped organisms were noticed.

Place where postmortem examination was done ....

Dated

Sd/ - Prasad

Signature of the autopsist

Designation

Postmortem No. 4

Species – Bovine

Colour –Gray

Owner –A. Roy

Residence – Changar,

Patna Breed – An indigenous cattle

Sex- Female

Age – About 4-year-old.

Weight – 500 kg.

By whom sent for examination and reasons, if any :

Veterinary Surgeon. To ascertain the cause of death.

Date and hour of death	Autopsy findings
10.12.93 am	1. Petechiae subepicardially
Date and hour of postmortem mucosae	2. Congested intestinal
Examination	
10.12.93 at 1.00 p.m.	3. Enlarged liver and spleen
	4. Anaemia and emaciation

History :-

Several cattle and buffaloes died within 2-3 hours with rise of temperature upto 1040 F. Anaemia and enlargement of the lymph nodes were noted in some ailing animals. Other symptoms noticed in the animals in the affected areas showed staggering gait, circling movements, frequent micturition, nervous excitement with attempts to press their heads against walls or fixed objects, Twitching of the muscles and profuse salivation were also seen.

Cadaverous changes – Pale conjunctivae. Poor nutritional condition (emaciated)

Rigor mortis – Present in the hind limbs.

External appearance The muscles were dry and reduced in

On removal of skin bulk.

(describe contusions and wounds etc.)

Mouth and pharynx – Pale mucosae of the mouth. A few petechiae near the junction of the lip mucous membrane with the skin.

Nasal cavities – N.A.D.

Larynx and trachea – N.A.D

Oesophagus -

Petechiae on the pleural surface. ½ Ib of serous fluid in the thorax.

Pleural cavity and lungs

Pericardium and heart (absolute wt.......... relative weight ....%) . A few subepicardial petechiae. Partially clotted blood in the left ventricle.

Peritoneum – About 2 Ibs of serosanguinous fluid in the peritoneal cavity.

Liver (absolute wt........3 kg. relative wt. 0.6%) Pale and moderately enlarged.

Gall bladder –N.A.D.

Spleen – Moderately enlarged (absolute wt. 1.5 kg relative wt. 0.3%)

Stomach and intestine – Abomasal and intestinal mucosae

Urinary organs were reddened (congested)

Kidneys- Slightly congested. (absolute wt........ relative wt. %)

Generative organs- N.A.D

Brain and spinal cord – Not opened.

Lymph glands in general – Slightly congested and prescapular lymph nodes enlarged.

Blood or tissue smears	1. Smears from the heart blood
Sent for microscopical	2. Smears from the spleen.

Examination

Diagnosis on the basis of

Above examination – **Surra.**

Remark (state if viscera and tissue specimens etc. sent for chemical, bacteriological, histopathological and parasitological examinations).

Microscopical examination of blood etc.

1 and 2 flagellated protozoa indistinguishable from *Trypanosoma evansi* were noticed.

Results of chemical, bacteriological

Parasitological and histopathological examinations.

Place where postmortem examination was done……………..

Dated :-….

Sd/ -B. Prasad

Signature of autopsist

Designation

Note: Blood films are preferably prepared during a parroxysm in the cases of Surra patients.

Postmortem No. -5

Species – Bovine

Colour – Grayish Red

Owner- v. Srinivasan

Breed – A crossbred cattle

Residence – Patna.

Sex – Female

Age- About 5-year-old of age

Weight – 600 kg.

By whom sent for examination and reasons. If any:

Veterinary Assistant Surgeon

To ascertain the cause of death.

Date and hour of death. Autopsy findings

10.12.60 at 10.00 a.m

Dated and hour of P.M. Exam.
10.12.60 at 10.00 a.m

1. Grayish white deposits and Erosions on the tongue and baccal mucous membrane
2. Haemorrhagic gastroenteritis
3. Subendocardial haemorrhages

History :-

Several cattle and buffaloes died with the symptom of high fever (105^0 F to 107^0 F) and erosive lesions on the tongue. The animals also showed thick creamy ocular discharge, dyspnoea, cough and dehydration. Erosions were noticed on the tongue and lower lip beneath the bran like deposits or raised grayish materials. Only one dead animal was carted here for autopsy.

Cadaverous changes – Blood tinged faucal matter at the anus. Conjunctive reddened with thick mucoid, discharge at the inner canthus Presence of ulcers on the dorsum of tongue.

Rigor mortis – Present in the hind limbs.

External appearance - on removal of skin (describe contusions and wounds etc.) – The muscles were dry and sticky to the touch.

Mouth and pharynx - Grayish brown deposits on the tongue and mucosae of the lips. When these deposits were removed. erosions or ulcers were visible.

Nasal cavities - N. A. D

Larynx and trachea – N.A. D

Pleural cavity and lungs- Both lungs were swollen. reddened and some what consolidated in pathches.

Pericardium and heart – Subendocardial and subepicardial haemorrhages. Presence of clotted blood in the left ventricle.

(absolute wt ....relative weight ....%)

Peritoneum – About 1 Ib of serous fluid in the peritoneal cavity)

Liver swollen and slightly pale. (absolute wt .... Relative wt ....%)

Gall bladder – Thick ropy bile.

Spleen- N.A.D (absolute wt ....relative wt................%)

Stomach and intestine – Abomasal, duodenal and intestinal mucosae reddened and covered with blood tinged mucoid material. Bands of congestion and haemorrhages in the rectal mucosae.

Urinary organs-

Kidneys – Swollen and pale.

(Absolute wt............../relative wt..................%)

Generative organs- N.A.D.

Brain and spinal cord- Not opened

Lymph glands in general – Presence of congestion (reddening) oedema and haemorrhages in the lymph nodes.

Blood or tissue smears sent -	1. Smears from the lymph nodes.
For microscopical examination -	2. Smears from the liver

Diagnosis on the basis of above examination **Rinderpest.**

Remarks (state if viscera and tissue specimens etc. sent for chemical, bacteriological, histopathological and parasitological examinations.)	(A) Pieces of lymph nodes and spleen in the sterile Petridishes

Microscopical examination of blood etc.

1, 2, A few G+ Putrefactive organisms

Results of chemical, Bacteriological, parasitological and histopathological Examinations	(A). A challenge of the two healthy and two immune calves with suspected splenic emulsion led to death of non immunised animals. Report on viral isolation awaited.

Place where postmortem examination was done ....

Dated:

K. Mohan

Sd/-

Signature of the autopsist

Designation

Postmortem No. -6

Species – Bovine

Colour – Black with white patches

Owner- B.K Sinha

Residence – Patna.

Breed – A crossbred cattle

Sex – Female

Age- About 4-year-old

Weight – 500 kg.

By whom sent for examination and reasons. If any:

Veterinary Assistant Surgeon

To ascertain the cause of death.

Date and hour of death

5.3.92 at 5.00 a.m.

Date and hour of P.M. Examination

5.3.92 at 9.00 a.m.

Autopsy findings:

1. Reddened oedematous prescapular lymph nodes
2. Abomasal ulcers
3. Reddened intestinal mucosae
4. Haemorrhages in the peritoneum and thorax.

History

Several exotic Jersey cattle and some crossbred animals showed enlarged superficial lymph nodes (e. g. suprascapular lymph nodes), rise of temperature (104^0 F to 106^0 F), rapid pulse, red conjunctivae, weakness, trembling. Cough, lachrymation, grinding of the teeth and some oedema in the throat region. Some sick Jersey cattle also showed symptoms of dysentery with loose blood tinged faeces. Marked fall in the milk yield was noticed.

Cadaverous changes – Pale conjunctivae. Superficial lymph nodes enlarged.

Rigor mortis- Rigor mortis present in the fore and hind limbs.

External appearance on removal of skin (describe contusions and wounds etc.) - Enlargement of prescapular & supramammary lymph nodes.

Mouth and pharynx- Presence of ulcers in the tonsillar regions Nasal cavities

Larynx and trachea- Mucosae in these parts were reddened.

Oesophagus- N.A.D

Pleural cavity and lungs-Swollen and oedematous. Blood tinged fluid escaped from the cut surfaces. Petechiae on pleural serosae.

Peritoneum - Peritoneal serosae studded with haemorrhages.

Liver- About one Ib of serosanguinous fluid in the peritoneal cavity.

(absolute wt................ relative wt...................%)

Swollen, enlarged and yellowish cut surfaces.

Gall bladder- slightly thick bile.

Spleen - N.A.D (absolute wt........................ relative wt ....%)

Stomach and intestine- Several oval or irregular shaped ulcers in the swollen and congested abomasal mucosae. The intestinal mucous membrane congested and covered with blood tinged mucoid material.

Urinary organs -

Kidneys (absolute wt...............relative wt...................%) The kidneys were swollen and pale.

Generative organs - N.A.D

Brain and spinal cord - Not opened

Lymph glands in general - Enlarged, oedematous and the cut surfaces were covered with blood tinged fluid.

Blood or tissue 1. Smears from the supramammary and prescapula lymph nodes.

Smears sent for microscopical

examination 2. Smears from the heart blood.

Diagnosis on the basis of above examination Theileriasis

Remarks (state if viscera and tissue specimens etc. sent for chemical, and parasitological examinations).

Microscopical examination of blood etc.

1. Faint blue masses containing red dots indistinguishable from Koch's blue bodies were noticed in the smears stained by Leishman's method.

Results of chemical, bacteriological parasitological and histopathological examination .

Place where postmortem examination was done ....

Date

Sd/- A.P. Roy

Signature of the autopsist

Designation

Postmortem No. -7

Species – Bovine

Colour – Grayish Red

Owner- D. Ram Breed – A crossbred bred jersey

Residence – Patna. Sex – Male

Age- About 6-year-old of age

Weight – 500 kg.

By whom sent for examination and reasons. If any: Resident

Surgeon Veterinary Hospital. To ascertain the cause of death.

Date and hour of death Autopsy findings-

10.12.65 at 3.00 a.m. 1. Yellowish discoloration
Date and hour of postmortem of the conjunctiva, aortic
exam. nitima and adipose tissue.

9.12.65 at 10.A.M.

2. Red colored urine.

3. Splenic enlargement

4. Reddened abomasal and intestina mucosae.

History :

The sick cattle showed symptoms of high fever (106^{0}F), anaemia, anorexia and rapid pulse and respiration. The conjunctivae and mucous membranes were brick red in the beginning but, later on changed to an yellowish discoloration. Incoordination and shivering were also noticed. The urine voided by the sick animal was dark red in colour. Red urine was also noticed in several favisian cattle brought to Bihar from Punjab with a history of high fever.

Cadeveverous changes Pale conjunctivae

Rigor mortis- Present in the fore limbs and jaw

External appearance on removal of skin (describe contusions and wounds etc.) The muscles were pale and sticky to touch. Thin and watery blood. Moderately oedematous subcutaneous tissue.

Mouth and pharynx – Presence of yellowish mucosae.

Nasal cavities- N.A.D.

Larynx and trachea- Pale mucosae

Oesophagus – Its mucosa showed yellowish tinge.

Pleural cavity and lungs - tinged Enlarged and oedematous. Blood fluid on the cut surfaces.

Pericardium and heart (absolute wt.....................relative wt......................%)

Partially clotted blood in the ventricles. Yellowish discoloration of the aortic intima. Mesenteric fat yellow.

Liver (absolute weight 3.5 relative wt. 0.7%)

Enlarged with yellowish brown cut surfaces.

Gall bladder - Thick ropy bile. About one Ib of serous fluid in the abdomen.

Spleen - Enlarged (4 times the normal size) (absolute wt. 2 kg relative weight - 0.4 %)

Splenic corpuscles prominently visible.

Stomach and intestine - Abomasal, duodenal and intestinal mucosae congested and swollen and covered with blood tinged ingesta.

Urinary organs-

Kidneys (absolute wt.................relative wt................... %). The kidneys are swollen with pale cut surfaces.

Generative organs - N. A. D

Brain and spinal cord - Not opened.

Lymph glands in general - Prescapular lymph nodes pale.

Blood or tissue smears sent for microscopical examination. 1. Smears from the heart blood.

2. Smears from the spleen

3. Two blood films prepared before death of the animal.

Diagnosis on the basis of above examination Babesiosis (Lal pesab)

Remarks (state if viscera and tissue specimens etc. sent for chemical, bacteriological, histopathological and parasitological examinations.)

Results of chemical, 1. Negative

2. and 3- Pear shaped forms of

Bacteriological *Babesia* species were noticed parasitopathological examinations.

Place where postmortem examination was done ....

Dated:

Sd/- A. Prasad

Signature of the Autopsist

Designation

Postmortem No. -8

Species – Bovine

Colour – Black with white patches

Owner- B. Singh

Breed – A crossbred cow

Residence – ....

Sex – Female

Age- About 5-year-old cow

Weight – 600 kg.

By whom sent for examination and reasons. If any: Resident surgeon,

Veterinary Hospital

To ascertain the cause of death

Date and hour of death

10.3.92 at 8.00 a.m.

Dated and hour of postmortem

Examination

10.3.92 at 11.00 a.m.

Autopsy findings

1. Pale conjunctivae
2. Thin and watery blood
3. Severely enlarged spleen
4. Gastroenteritis

History :

A crossbred male calf showed symptoms of emaciation, anaemia, icterus, rise of temperature (105^{0}F). Rapid muscle tremors and difficult movement were noticed in the sick calf.

Cadaverous changes-Conjunctivae of the eyes were pale.

Rigor mortis-Present in the head and limbs.

External appearance on On removal of skin (describe Contusions and wound etc.) Pale muscles and sticky to touch. Blood was thin and watery.

Mouth and pharynx- Pale mucosae of these cavities

Nasal cavities- N. A. D

Larynx and trachea – Their mucosae were pale

Oesophagus- N. A. D

Pleural cavity and lungs

Pericardium and heart (absolute wt................ relative wt....................%)

Subepicardial petechiae. Clotted blood in the left ventricle of heart .

Peritoneum- About on Ib of serous fluid in the peritoneal cavity.

Liver(absolute wt.................... relative wt.....................%)

Enlarged with yellowish cut surfaces.

Gall bladder – it contained dark thick bile.

Spleen (absolute wt................2 kg. relative wt....................33%)

Severely enlarged with prominent splenic corpuscle and splenic pulp was reddish.

Stomach and intestine- Abomasal and intestinal mucous membrane congested and covered with blood tinged ingesta.

Urinary organs – kidneys (absolute wt.............relative wt..............%)

- These were enlarged and showed the pale cut surfaces)

Generative organs- N. A. D

Brain and spinal cord- Not opened.

Lymph glands in general – Pale.

Blood or tissue smears sent	1. Blood smears from the heart.
For microscopical examination	2. Smears from the spleen.

Diagnosis on the basis of above examination - **Anaplasmosis**

Remarks (state if viscera and Tissue specimens etc.	A. Two blood films prepared before death of the animal.

Sent for chemical, bacteriological, histopathological and parasitological examinations)

Microscopical examination of blood etc.

results of chemical, bacteriological parasitological examination.	1. Anaplasma detected in the stained film by Leishman's method. The organisms appeared in the red cells as oval or round red stained dots. **indistinguishable from** *Anaplasma* **spp.**

Place where postmortem examination was done...................

Date:

Sd/- A. K. Sinha

Signature of the autopsist

Designation

Postmortem No. -9

Species – Bovine

Colour – Red

Owner- D. Prasad
Breed – Haryana

Residence – Patna.
Sex – Female

Age- About 5-year-old cow
Weight – ....

By whom sent for examination and reasons. If any:

Veterinary Surgeon

To ascertain the cause of death.

Date and hour of death	Autopsy findings
10.1.1966 at 2.00 a. m.	1. An open wound in the thigh
Date and hour of postmortem Examination. 10.1.1966 at 9.00 a.m	2. Accumulated pus in the uterine cavity. 3. Congestion in the medulla and spinal cord.

History :

A cow with a recent history of calving was suffering from purulent endometrits after a week of calving. Tetany convulsions, rigidity of muscles, difficult swallowing, labored respiration, and tympanitis were noticed in the ailing cow. Penicillin and terramycin were administered but the animal could not recover from its illness.

Cadaverous change – An open wound infested with maggots was noticed in the thigh muscles.

Rigor mortis The muscles were dry and sticky to touch.

Removel of skin (describe contusions and wounds etc.)

Mouth and pharynx –N.A. D

Nasal cavities – N. A. D

Larynx and trachea- N. A. D

Oesophagus- N. A. D

Pleural cavity and lungs- N. A. D

Pericardium and heart (absolute wt........................relative wt.................. %)

Partially clotted blood in the left ventricle.

Peritoneum – N. A. D

Liver (absolute wt.........................relative wt............... %)

Slightly congested

Gall bladder – N. A. D

Spleen – Pale (absolute wt......................relative wt.................... %) –N. A. D

Generative organs- About 2 litres of purulent exudates (pus) in the uterus. The endometerium showed grayish white necrotic patches.

Brain and spinal cord- Opened There was slight congestion in the spinal cord and medulla.

Lymph glands in general – N.A. D

Blood or tissue smears sent

For microscopical examination

1. Smears from the wounds in the thigh
2. Smears from the heart blood
3. Smears from the uterine exudate

Diagnosis on the basis of above examination.

Tetanus (Lock jaw)

Remark's (state if viscera and tissue specimens etc, sent for Chemical bacteriological, histopathological and parasitological examination)

1. Samples of pus in a sterile pipettes were collected from the wounds in the thigh and uterine mucosa for bacteriological investigation.

Microscopical examination of blood etc.

1, 2, and 3 – A few Gram positive rods. Terminal spores in the rods were noticed.

Results of chemical, bacteriological,

Parasitological and histopathological examinations.

Place where postmortem examination was done....................

Date:

Note: A presumptive diagnosis of tetanus was given on the basis of symptoms like respiratory distress, convulsion, asphyxia and presence of wound in the thigh and congestion in the spinal cord.

Sd/- A. Prasad.

Signature of the autopsist

Designation

Postmortem no. -10

Species – Bovine

Colour – Gray

Owner- Cattle Farm — Breed – Tharparkar

Residence – Patna. — Sex – Female

Age- About 6 – year - old cow

Weight – ....

By whom sent for examination and reasons. If any:

Veterinary Assistant Surgeon

To ascertain the cause of death.

Date and hour of death	Autopsy findings
5.7.1969 at 2.00 a. m	1. Grayish white foci in the endometerium
Date and hour of P.M exam.	and foetal cotylendons
5.7.1960 at 9 a.m	2. Petechiae on the serosae and endocardium

History:

Many cows aborted after 5-6 months of pregnancy. Placentae of foetal membranes were frequently retained in these cows. Slightly brownish discharge from the vulvar orifice of cows. The aborted foetus and its dead dam were received here for postmortem examination.

Cadaverous changes- Conjunctivae of the eyes were reddened. Blood tinged thick discharge from the vulva.

Rigor mortis- Present in the limbs.

External appearance on	Muscles were dry and sticky to touch.

removal of skin (describe contusions and wounds etc.)
Mouth and pharynx - N. A. D
Nasal cavities- N. A. D
Larynx and trachea- Had pale mucosae
Oesophagus- N. A. D
Pleural cavity and lungs
Pericardium and heart - Subepicardial and subendocardial haemorrhages.

(absolute wt...................relative weight ....%)

Partially clotted blood in ventricles of the heart. Right ventricle dilated with a few subendocardial grayish areas.

Peritoneum - A few petechiae on the peritoneum.

Liver -Swollen and congested, (absolute wt..................... relative wt............. %)

Gall bladder – It contained thick bile.

Spleen – N. A. D (absolute wt............relative wt........... %)

Stomach and intestine – Distended with gas.

Urinary organs – N. A. D

Kidneys (absolute wt ....relative wt....................%)

Generative organs

Uterus- Endometerium thickened with grayish white foci. Foetal cotyledons presented a leathery and grayish white patchy areas. Thickening of the chorium.

Brain and spinal cord – Not opened.

Lymph glands in general – Supramammary lymph nodes were enlarged and swollen.

Blood or tissue smears sent for	1. Smears from the endometerium.
Microscopical examination -	2. Smears from the stomach contents of the aborted foetus.

Diagnosis on the basis of above

Examination- Brucellosis complicated with toxaemia.

Remarks (state if viscera and tissue specimens etc. sent for chemical bacteriological and parasitological Examinations)	A. Stomach contents aspirated into sterile pipettes (two) were collected and sent to the Bacteriology department for isolation.

Microscopical examination of blood etc.

1, 2-A few Gram negative rods.

Results of chemical, bacteriological

Parasitological and histopathological Examination.	A. *Brucella abortus* Isolated.

Place where postmortem examination was done ……………..

Date:

Sd/- B. Sen

Signature of the autopsist

Designation

Postmortem No. -11

Species – Bovine

Colour – Red

Owner- B. Roy

Residence – Patna.

Breed – A crossbred bred cattle

Sex – Female

Age- 2 month -old calf

Weight – 40 kg.

By whom sent for examination and reasons. If any:

Veterinary Surgeon

To ascertain the cause of death.

Date and hour of death/ sacrifice

5. 1. 1993 At a. m.

date and hour of P.M. exam.

5. 1. 1993 at 9 a.m

Autopsy findings.

1. Bronchopneumonia as a terminal disease.
2. Ulcers on the feet, tongue and lips.

History :

Several cross - bred calves in a farm showed lameness, rise of temperature (104^{0}F), anorexia, depression and salivation with long ropy strings. Vesicles and ulcers were noticed on the buccal mucosa, tongue and dental pad. Ulcerative foot lesions were also seen in the skin of the foot (inter digital cleft). An emaciated calf with foot and mouth ulcerative lesions was received for autopsy.

Cadaverous changes : Pale conjunctivae . Ulcers and vesicles on the tongue and skin on the coronary band. The feet, interdigital cleft and nostrils also revealed vesicles and ulcers.

Rigor mortis : Present in the hind legs. A few ulcers on the heels.

External appearance on removal- of skin (describe contusions and wounds etc.)	Muscles dry and sticky to touch.

Mouth and pharynx – Ulcers on the dorsum of the tongue.

Nasal cavities- N.A. D

Larynx and trachea- Congestion in the mucosae of the larynx and trachea.

Oesophagus- N. A. D

Pleural cavity and lungs, the lungs were congested (reddened) and consolidated (firm) in consistency in patches and small consolidated cat pieces sank in water.

Pericardium – Subepicardial and subendocardial haemorrhages and heart (absolute wt................relative weight ....%)

Grayish white steaks in the cardiac musculature.

Peritoneum – A few haemorrhages on the peritoneal serosae.

Liver (absolute wt................. relative wt.................%)

Enlarged and moderately congested.

Spleen – pale (absolute wt................. relative wt...........%)

Stomach and intestine – Abomasal and intestinal mucosae reddened and covered with blood tinged ingesta.

Urinary organs-

Kidneys (absolute wt...................relative wt................%)

Swollen and pale.

Generative organs- N. A. D.

Brain and spinal cord- Not opened.

Lymph glands in general - Pale

Blood or tissue smears sent for- Microscopical examination

1. Smears from the heart.
2. Smears from the lungs.

Diagnosis on the basis of above examination.

Foot and mouth disease complicated with bronchopneumonia.

Remarks (state if viscera and tissue specimens etc. sent for chemical, bacteriological histopathological and parasitological examination)

Microscopical examination of blood etc.

1. A few Gram positive nods.

Results of chemical, bacteriological parasitological and histopathological examinations.

Place where postmortem examination was done ....

Dated:

Note:

An outbreak of F. M. D. involving young calves was seen in a Govt. farm. Greyish areas in the heart musculature (indicating degeneration) were noticed.

Seena

Sd/-

Signature of the autopsist

Designation

Postmortem No. -12

Species - Canine

Colour - Black with White

Owner- Anup Ram

Breed - A street dog (nondescript)

Residence - Patna.

Sex - Male

Age- About 6-Year-old dog
Weight – ....

By whom sent for examination and reasons. If any:

Veterinary Assistant Surgeon

To ascertain the cause of death

Date and hour of death/sacrifice	Autopsy findings
4. 3. 90 at 10.00 a.m	1.Broken teeth
Date and hour of P.M exam.	2. Worn out pads
4. 3. 90 at 2. 00	3. Meningeal congestion
	4. Negri bodies in the impression smears of brain

Histroy: As reported by the owner, his pet was bitten by a street dog 8 (eight) weeks ago and reported to lie down somewhere in corner of rooms. The symptoms showed by his pet included profuse salvation, protrusion of the tongue, inability to drink milk or water and dropped jaw. It showed high rise of temperature (105^0 F) after the bite by the street dog. But, later on, it ran subnormal temperature at the time of admission into the hospital and died on the 4th day after admission in the dog ward

Cadaverous changes: Emaciated carcase. Reddened conjunctivae of the eyes. A few abrasions on the right thoracic wall. An open wound on the lateral surface of left thigh.

Rigor mortis: Present in the hind limbs. Lacerations in the ears. Foot pad was worn out.

External appearance on removal of

Skin (describe contusions and wounds etc.)

Mouth and pharynx	–	Broken canine teeth.
Nasal cavities		Not opened
Larynx and trachea		Do

Oesophagus Do

Pleural cavity and lungs Do

Pericardium and heart(absolute wt....................relative wt....................%)

Peritoneum- Do

Liver (absolute wt..................relative wt ....%) Do

Gall bladder- Do

Spleen (absolute wt...........................relative wt..................%)

Not opened

Stomach and intestine –Do

Urinary organs- Do

Kidneys (absolute wt...................relative wt...........................%)

Generative organs –Not opened.

Brain and spinal cord – The brain was removed and impression smears were prepared from the hippocampus major. The meaninges were moderately congested.

Lymph glands in general – Not opened.

Blood or tissue smears sent for Examination — 1. Impression smears from the hippocampus major of the brain

Diagnosis on the basis of above examination – **Rabies**

Remarks (state if viscera and tissue specimens etc. sent for chemical, bacteriological, histopathological and parasitological examinations.) A brain (hippocampus) preserved in 50% glycerine saline for biological test.

Microscopical examination of blood etc.

1. Inclusion bodies indistinguishable from Negri-bodies were detected in the smears stained by Seller's technique.

Results of chemical, bacteriological parasitological and histopathological examinations. A Virological report awaited.

Place where postmortem examination was done....................

Dated:

Noted:

When Negri bodies are not detected in the impression smears of the suspected cases of rabies, the carcase is opened to find out the cause of death. But the autopsist has to mention in his report that a mere absence of Negri bodies does no absolutely exclude the possibility of rabies. A report of this kind enables the clinician to suggest vaccination against Rabies to prevent its occurrence in future. It is safe for humans to consult medical Doctors for protection against rabies.

Sd/-

K. Mohan

Signature of the autopsist

Designation

Postmortem No. -13

Species – Canine

Colour – Black with grayish abdominal patches.

Owner- Police dog. Breed – An Alsatian crossbred bred dog

Residence – Patna. Sex – Male

Age- About 5 -Year- old dog

Weight – ....

By whom sent for examination and reasons. If any Veterinary Assistant Surgeon. To ascertain the cause of death

Date and hour of death/ sacrifice Autopsy findings

1.6. 1980 at 10 a.m

date and hour of p.m. exam.
1.6.1980 at 3 P.M.

1. Haemorrhagic inflammation in the stomach and intestines.

2. Faecal matter mixed with blood tinged material

3. Dehydration.

History:

Deaths of several police dogs at Patna with history of bloody diarrhea, anaemia and dehydration and vomiting at times were reported.

Cadaverous changes – Pale conjunctivae.

Rigor mortis – Present in the jaw, fore limbs and hind limbs.

External appearance on – Muscle, pale, dry and sticky to touch.

Removal of skin (describe

Contusions and wounds etc)

Mouth and pharynx – Pale

Nasal cavities – N. A. D.

Larynx and trachea – Congested mucosae.

Oesophagus – N. A. D.

Plural cavity and lungs – Both the lungs were congested and blood escaped from the cut surfaces.

Pericardium and heart (absolute wt ....relative wt...............%)

Subepicardial and subendocardial haemorrhages.

Clotted blood in the left ventricle .

Peritoneum

Liver – Congested (absolute wt ....relative wt................%)

Gall bladder – It had thick bile.

Spleen – Moderately enlarged. (Absolute wt............relative wt............%)

Stomach and intestine –The mucosae of the stomach, and intestines were congested, Thickened, haemorrhagic and covered with bloody contents.

Urinary organs

Kidneys – Pale and swollen (absolute wt.............relative wt..............%)

Generative organs – N.A. D.

Brain and spinal cord – Not opened

Lymph glands in general mesentric lymph nodes swollen and reddened

Blood escaped from the cut surfaces.

Pericardium and heart (absolute wt.........................relative wt.................%)

Subepicardial and subendocardial haemorrhages.

Clotted blood in the left ventricle.

Peritoneum

Liver – Congested (absolute wt.........................relative wt. ....%)

Gall bladder – Had presence of thick bile.

Spleen – Moderately enlarged (absolute wt.................relative wt................%)

Stomach and intestine –The mucosae of the stomach, and

intestines congested, thickened, haemorrhagic and covered with bloody contents.

Urinary organs

Kidneys – Pale and swollen (absolute wt ....relative wt................%)

Generative organs – N. A. D.

Brain and spinal cord – Not opened

Lymph glands in general Mesentric lymph nodes swollen and reddened

Blood or tissue smears sent for – microscopical examination	1. Smears from the heart. 2. Examination of blood tinged intestinal contents. 3. Smears from spleen of the dog.
Diagnosis on the basis of above exam.	Haemorrhagic gastroenteritis

(Canine Parvovirus infection)

Remarks (state if viscera and tissue specimens etc. sent for chemical, bacteriological histopathological and parasitological examinations)

A. Portions of liver, spleen and intestines collected for viral isolation.

B. Small pieces preserved in 10% formalin solution for histopathology .

Microscopical examination of blood etc.

1, 3 – Nagetive

2 – Negative

Results of chemical, bacteriolical

Parasitological and histopathological

Examinations A & B -Reports not received

Place where postmortem examination was done ....

Dated:

Note: A presumptive diagnosis of haemorrhagic gastroenteritis was given subject to confirmation by the parvovirus isolation in the laboratory.

S. P. Rama

Sd/-

Signature of the autopsist

Designation

Postmortem no. -14

Species – Bovine

Colour – Grey

Owner- G. Prasad

Residence – Patna.

Breed – Tharparkar

Sex – Female

Age- About 2-year month -old cow

Weight – 800 kg.

By whom sent for examination and reasons. If any:

Veterinary Surgeon, Phulwari, Patna.

To ascertain the cause of death.

Date and hour of death/ sacrifice — Autopsy findings

10.12.94 at 3.00 a.m. — 1. Caseated nodules in the lungs.

Date and hour of P.M. exam. — 2. Caseated bronchial and media stinal lymph nodes.

9. 12. 96 at 10. 00 a.m. 3. Pink jelly like material in the coronary groove of heart

History:

An emaciated carcase of a cow was received for autopsy. It had rough hair coat with history of capricious appetite, chronic cough and fluctuating temperature. The bony eminences were prominent and the conjunctivae of the eyes were pale.

Cadaverous changes: Pale conjunctivae of the eyes. Prominent bony eminences. Visible thoracic ribs. (Hide bound condition)

Rigor mortis- Present in the jaw and fore limbs.

External appearance on removal of – Subcutaneous fat scanty and gelatinous at some places.

Skin (describe contusions and wound etc.) Muscles pale and reduced in bulk.

Mouth and pharynx – N. A. D.

Nasal cavities – N. A. D.

Larynx and trachea – Had pale mucosae.

Oesophagus – N.A.D.

Pleural cavity and lungs – Several greyish white nodules in the lung parenchyma. When cut, the nodules showed yellowish grey caseous surfaces. Caseated cream colored cut surfaces of the enlarged bronchial and mediastinal lymph nodes were seen. A characteristic obnoxious smell was emitted from the cut surfaces of the lymph notes

Pericardium and heart (absolute wt....................relative wt..............%)

Pink jelly like material in the coronary grove and clotted blood in the left ventricle of heart.

A Ib of serosanguinous fluid.

Peritoneum – N. A. D.

Liver (absolute wt...............relative wt........................%)

Pale cut surfaces A few abscesses on the cut surfaces of the livers.

Gall bladder – Thick ropy bile.

Spleen – Reduced in bulk with wrinkled capsule.

(Absolute wt.....................relative wt ....%)

Stomach and intestine- Partly disteneded with gas.

Urinary organs-

Kidneys (absolute wt................relative wt...................%)

Swollen and partly congested.

Generative organs – N.A.D.

Brain and spinal cord – Not opened.

Lymph glands in general – Submandibular lymph nodes enlarged and had cream colored caseated cut surfaces.

Blood or tissue smears sent for	1. Smears from the cut surfaces of the microscopical examination pulmonary nodules
Diagnosis on the basis of above	2. Smears from the cut surfaces of the examination **Tuberculosis (TB)** mediastinal lymph nodes.

Remarks (state if viscera and tissue specimens etc. sent for chemical, bacteriological, histopathological and parasitological examination)

Microscopical examination of blood etc.

1 and 2. Acid fast rods indistinguishable from ***Mycobacterium tuberculosis*** in the smears stained by **Ziehl Nielsen's** technique.

Results of chemical, bacteriological, parasitological and histopathological examinations.

Place where postmortem examination was done..................

Dated:

Sd/- B. Ram

Signature of the autopsist

Designation

Postmortem No. -15

Species – Bovine

Colour – Grey

Owner - Cattle

Farm, Patna

Residence – Patna.

Breed – Tharparkar

Sex – Female

Age- About 8-year month -old cow

Weight – 500 kg.

By whom sent for examination and reasons. If any:

Farm Veterinary Assistant Surgeon

To ascertain the cause of death.

Date and hour of death/sacrifice	Autopsy findings
6. 4. 1985 at 11.00 a. m.	1. Thickened folds into the intestinal mucosae.
Date and hour P.M. exam.	2. Enlarged and swollen mesenteric lymph nodes
	3. Jelly like material in the coronary groove of heart.

4. Thin and watery blood.

History:

An emaciated carcase with a history of persistent or recurrent diarrhea was received. The conjunctivae of the cow were pale and the bony eminences were very prominent with rough coat and staring ribs. The animal was eating as usual. Faeces was soft and at times, thin like soup.

Cadaverous changes – Pale conjunctivae of the eyes. Thoracic ribs and bony eminences prominent Hide bound staring coat and anal region soiled with loose faecal matter.

Rigor mortis- Present in the hind limbs.

External appearance on removal of skin (describe contusions and wounds etc.)

Pale muscle and subcutaneous fat gelatinised. Blood was thin and watery. Muscles dry and sticky to touch.

Mouth and pharynx- Had pale mucosae

Nasal cavities – N. A. D.

Larynx and Trachea- had pale mucosae

Oesophagus – N. A. D.

Pleural cavity and lungs – Both the lungs were pale. Presence of jelly like material in the coronary groove of heart.

Pericardium and heart (absolute wt................relative wt................%)

Partly clotted blood in the left ventricle of heart, Pale cardiac musculature.

Peritoneum – About 2 Ib of serosanguinous fluid in the abdomen.

Liver- Pale. (absolute wt..........................relative wt.........................%)

Gall bladder – Had thick ropy bile.

Spleen – Pale with wrinkled capsule. Reduced in volume

(absolute wt................relative ....%)

Stomach and intestine – Ingesta in the fore stomachs Intestinal mucous membrane thickened, grey, and corrugated in appearance. Mucosae near the ileocaecal valve were also thrown into folds.

Urinary organs- N. A. D.

Kidneys (absolute wt ....relative wt..................%)

Pale

Generative organs – N. A. D.

Brain and spinal cord – Not opened.

Lymph glands in general – Mesenteric lymph nodes swollen, oedematous and pale.

Blood or tissue smears sent for microscopical examination	1. Smears from the intestinal mucous membrane
	2. Smears from the mesenteric lymph nodes.

Diagnosis on the basis of above examination – **Johne's disease (JD)**

Remarks (state if viscera and tissue specimens etc. sent for chemical, bacteriological histopathological and parasitological examination)

Microscopical examination of blood etc.

1, 2, Acid-fast rods indistinguishable from

Mycobacterium Paratuberculosis in the smears

stained by Ziehl-Nelson's method.

Results of chemical, bacteriological

Parasitological histopathological

examinations

Place where postmortem examination was done.................

Date -

Sd/- P. N. Sinha

Signature of the autopsist

Designation

Postmortem No. -16

Species – Equine

Colour – Grey

Owner- P. Sinha

Residence – Bihta

Breed – A local indigenous horse
Sex – Male

Age- About 4-year-old horse
Weight – ....

By whom sent for examination and reasons. If any:

Veterinary Assistant Surgeon,Bihta.

To ascertain the cause of death.

10.12.62 at 9.00 a. m.

Date and hour of P. M. exam.

10.12.62 at 11 a.m.

1. Subcutaneous oedema in the head, neck, sternum and absence of natural wrinklings.
2. Subendocardial haemorrhages and hydropericardium
3. Excess of fluid in the abdomen

History :

The affected horse showed a rise of temperature, restlessness and glaving oedematous swellings in the head and temporal fossa with swollen eye lids, Only one carcase was available for autopsy.

Cadaverous changes – Conjunctivae of the eye reddened. Mucoid discharge at the inner canthus. Swollen head and temporal fossa.

Rigor mortis - Present in the fore and hind limbs.

External appearance on removal of skin (describe contusions and wounds etc.)

Oedematous changes in the subcutaneous tissue of the neck and brisket.

Mouth and pharynx - N. A. D.

Nasal cavities - Congested mucosae.

Larynx and trachea - Congested mucosae in these parts

Oesophagus - N. A. D.

Pleural cavity and lungs

Pericardium and heart (absolute wt ....relative wt....................%) A pound of fluid in the paricardial cavity

Subepicardial and subendocardial haemorrhages. Clotted blood in the left ventricle.

Peritoneum - About 2 litres of fluid in the peritoneal cavity.

Liver- Swollen and congested

Gall bladder - It contained thick bile.

Peritoneum It contained a Ib of turbid fluid

Spleen - A few subcapsular petechiae (absolute wt ....relative wt..........%) stomach and intestine –N. A. D.

Urinary organs

Kidneys swollen and partly congested (absolute wt ....relative wt............%)

Generative organs - N. A. D.

Brain and spinal cord - Not opened

Lymph glands in general - N. A. D.

Blood or tissue smears sent for - 1. Smears from the heart.

Microscopical examination 2. Smears from the lungs.

Diagnosis on the basis of above examination – African horse sickness.

Remarks (state if viscera and tissue

Specimens etc. sent for chemical,

Bacteriological, histophatholical and

parasitological examinations)

Microscoscopical examination of blood etc.

1 and 2 Negative

Results of chemical, bacteriological parasitological and histopathological examinations.

Place where postmortem examination was done ....

Dated :

Note :

A presumptive diagnosis based on the symptoms and gross lesions is given prior to receipt of the report on the bacteriological and virological investigation conducted in the laboratory.

Md. Ashraf

Sd/-

Signature of the autopsist

Designation

Postmortem no. -17

Species – Bovine

Colour – Cattle Farm,

Patna

Residence – Patna.

Breed – Tharparkar

Sex – Female

Age- About a-year month -old calf

Weight – ....

By whom sent for examination and reasons. If any:

Farm Veterinary Assistance Surgeon

Date and hour of death/sacrifice	Autopsy findings
12.8.60 at 3.00 a. m.	1. An open swollen navel wound
Date and hour of P. M. exam.	2. Liver abscesses
10/8/60 at 9.00 a. m.	3. Blood clot in the umbilical vein.

History:

A newly born female calf showed swelling in its navel with rise of temperature (103^0 F). There was some swelling in its fetlock joint. The calf was treated with antibiotics but in vain.

Cadaverous changes – Nutritional condition poor. Navel region swollen with blood tinged discharge.

Rigor mortis – Present in the jaw and facial.

External appearance on removal of skin (describe contusions and wounds etc)

Muscles pale and sticky to the touch. When the umbilical swelling was opened, blood tinged purulent exudate was noticed in it.

Mouth and pharynx – Presence of pale mucosae.

Nasal cavities – N. A. D.

Larynx and trachea – Pale mucosae.

Oesophagus – N. A. D.

Pleural cavity and lungs – Both the lungs were congested and consolidated in patches. The cut pieces of the consonlidated lungs (firm like liver) sank into water.

Pericardium and heart (absolute wt.......................relative wt...................%)

Partly clotted blood in the left ventricle. Pale cardiac musculature. A few fibrinous flakes on the mesentery.

Peritoneum – Blood tinged fluid in the abdominal cavity.

Liver (absolute wt ....relative wt......................%)

Swollen and congested. Had cavities containing cream colored purulent material (abscesses.)

Gall bladder – It contained thick bile. Umbilical vein contained poorly clotted blood. Portal lymph nodes swollen, oedematous and congested.

Spleen – Pale with wrinkled capsule (absolute wt............relative wt..............%)

Stomach and intestine – Abomasal and intestinal mucous membrane congested and swollen and contained blood tinged ingesta.

Urinary organs

Kidneys (absolute wt....................relative wt...............%)

Swollen and congested.

Generative organs – N.A.D,

Brain and spinal cord- Not opened

Lymph glands in general – Mesenteric lymph nodes moderately congested.

Blood or tissue smears sent for	1. Smears from the umbilical swelling
Microscopical examination	2. Smears from the lungs.

Diagnosis on the basis of above examination

Navel – ill with terminal bronchopneumonia.

Remarks (state if viscera and tissue specimens etc. sent for chemical, bacteriological, histopathological and parasitological examinations)

Microscopical examination of blood etc.

1. Gram negative nods and Gram positive cocci
2. Gram positive put refractive rods.

Results of chemical, bacteriological, parasitological and histopathological examination.

Place where postmortem examination was done ....

Dated –

Sd/- A. Prasad

Signature of the autopsist

Designation

Postmortem no. -18

Species – Bovine

Colour – Red

Owner- B. Ram

Residence – Patna.

Breed – A cross bred calf

Sex – Male

Age - 6 month -old – calf

Weight – 150 kg.

By whom sent for examination and reasons. If any:

Veterinary Surgeon, Phulwari, Patna.

To ascertain the cause of death.

Date and hour of death Autopsy findings

5. 2. 88 at 2.00 am — 1. Intestinal mucous membrane

Date and hour of p.m. exam. congested and Haemorrhagic
5.2.88 at 9.00 a.m. — 2. Anaemia and emaciation

History:

As per history, the calf showed tenesmus and was suffering from diarrhea, dysentery and was anaemic and emaciated.

Cadaverous changes- Anal region soiled with blood tinged faecal matter with foul smell. Poor nutritional condition-Anaemic carcase.

Rigor mortis- Present in the hind limbs. External appearance on removal of skin (describe contusions and wounds etc.)

Muscles dry and pale and sticky to the touch. Blood was thin and watery. Muscles reduced in bulk.

Mouth and pharynx – Presence of pale mucosae

Nasal cavities – N. A. D.

Larynx and trachea – Pale mucosae seen.

Oesophagus – N. A. D.

Pleural Cavity and Lungs- Pale

Pericardium and heart (Absolute Wt. ....Relative Wt. ....%)

Partly clotted blood in the left ventricle.

Peritoneum- A Ib of serous fluid in the abdomen.

Liver – N. A. D. (absolute wt...........relative wt...............%)

Gall bladder – It continued thick bile.

Spleen – Pale with wrinkled capsule (absolute wt.................. relative wt...........%)

Stomach and intestine- Intestinal mucous membrane congested, thickened and covered with blood thinged material.

Haemorrhages in the ileum,

Urinary organs

Kidneys - Pale (absolute wt...............relative wt...........%)

Generative organs - N. A. D.

Brain and spinal cord- Not opened

Lymph glands in general - Pale

Blood or tissue smears sent for microscopical examination

1. Examination of the mucoid faecal content
2. Smears from the heart blood.

Diagnosis on the basis of above examination - Coccidiosis

Remark (state if viscera and tissue specimens etc. sent for chemical, bacteriological, histopathological and parasitogical examinations.)

Microscopical examination of blood etc.

1. **Oocysts of *Eimeria* spp. detected.**

2. Negative

Results of chemical, bacteriological parasitological and histopathological examinations.

Place where postmortem examination was done.................

Dated:

Y. K. Jain

Sd/-

Signature of the autopsist

Designation

Postmortem no. -19

Species - Bovine

Colour - Red

Owner- B.B. Verma

Residence – Patna.

Breed – A cross bred calf

Sex – Female

Age- About 20 –day-old calf

Weight – 60 kg.

By whom sent for examination and reasons. If any:

Veterinary Surgeon, Phulwari, Patna.

To ascertain the cause of death.

Date and hour of death/ sacrifice	Autopsy findings
10.2.92 at 1.00 a. m.	1. Pale intestinal mucosae
Date and hour of postmortem	2. Tail and buttock soiled
Examination	with faeces.
10.12. 92 at 9.00 a. m.	3. Dry tongue

History:

As reported, the calf showed symptoms of depression, weakness, loss of appetite, rise of temperature (104^{0}F), diarrhea and dysentery. Its faecal matter was pasty and chalky white in colour. Tail and buttocks were soiled with rancid faecal matter, The calf was treated with antibiotics.

Cadaverous changes- Eyes sunken. Pale conjunctivae, skin loose, tongue dry and parched and clay colored obnoxious smelling faeces at the anus.

Rigor mortis – Present in the hind limbs

External appearance on removal of skin (describe contusions and wounds etc.) – Muscles dry and sticky to the touch.

Mouth and pharynx – Dry mucosae

Oesophagus – N. A. D.

Pleural cavity and lungs – Pale

Pericardium and heart (absolute wt.................... relative wt..............%) Clotted blood in the left ventricle.

Peritoneum – N. A. D.

Liver (absolute wt.................relative wt.......................%)

Gall bladder – It contained thick ropy bile

Spleen – It is swollen (absolute wt........................relative wt....................%)

Stomach and intestine- The abomasal and intestinal mucosae had pale appearance and covered with mucoid ingesta and congested mucosal appearance.

Kidneys – Congested (absolute wt.........................relative wt...................%)

Generative organs – N. A. D.

Blood and spinal cord – Not opened

Lymph glands in general – Mesenteric lymph nodes swollen and congested.

Blood or tissue smears sent for microscopical examination

1. Smears from the intestinal mucosa.
2. Smears from the cut surface of the liver.
3. Faecal swabs.

Diagnosis on the basis of above examination – **Colibacillosis** (White scour)

Remarks (state if viscera and tissue specimens etc. sent for chemical, bacteriological histopathological and parasitological examinations)

Microscopical examination of blood etc.

1, 2 – Few Gram negative rods.

3. E. coli were isolated.

Results of chemical, bacteriological parasitological and histopathological examinations.

Place where postmortem examination was done ....

Dated-

Sd/-

P. Lal

Signature of the autopsist

Designation

Postmortem no. -20

Species – Swine

Colour – Black

Owner- C. Prasad

Residence – Patna.

Breed – An indigenous one

Sex – Female

Age- About 1-year old swine

Weight – 40 kg.

By whom sent for examination and reasons. If any:

Veterinary Surgeon, Phulwari, Patna.

To ascertain the cause of death.

Date and hour of death/ sacrifice

10.12.65 at 3.00 a. m.

Autopsy findings.

1. Subendocardial haemorrhages

2. Gastroenteritis

Date and hour of

Postmortem examination

3. Subcapsular petechiae in the kidneys.

10.12.65 at 10.00 a. m.

History:

Several pigs died with high temperature and constipation followed by diarrhea. Other symptoms noticed in the pigs were yellow ocular discharge, foul smelling faecal matter mixed with blood, cough and difficult respiration. Most of the animals died within a day or two and one case was sent for autopsy.

Cadaverous changes – Conjunctivae of the eyes were reddened.

Rigor mortis – Present in the hind limbs, which could not be flexed.

External appearance on removal of skin (describe contusions and wound etc.)

Muscles dry and sticky to touch

Mouth and pharynx – N. A. D

Nasal cavities - N. A. D.

Larynx and trachea – A few petechiae in the tracheal mucosae

Oesophagus- N. A. D.

Pleural cavity and lungs – Both the lungs were swollen, firm and reddened in patches. A few petechiae on pleural serosae

Pericardium and heart (absolute wt..............relative wt........................%)

Subendocardial and subepicardial petechiae.

Peritoneum – Abort 1 Ib of serosanguinous fluid in the peritoneal cavity.

Liver (absolute wt.....................relative wt..................%)

Enlarged. Congested and a few subcapsular petechiae

Gall bladder: N. A. D.

Spleen – Enlarged with subcapsular haemorrhages

(Absolute wt.........................relative wt..........................%)

Stomach and intestine - Mucosae of the stomach and

intestine swollen, congested and covered with blood tinged mucoid ingesta, Haemorrhagic ulcerative intestinal lesions.

Urinary organs

Kidneys (absolute wt.............relative wt.......................%)

The kidneys were swollen and showed several subcapsular haemorrhages, Petechiae in the mucosae of urinary bladder.

Generative organs – N. A. D.

Brain and spinal cord – Not opened

Lymph glands in general – Moderately swollen and congested

Blood or tissue smears sent for Microscopical examination.
1. Smears from the lungs.
2. Smears from the heart
3. Smears from the kidneys

Diagnosis on the basis of above examination – **Swine fever (Hog cholera)**

Remarks (state if viscera and tissue specimens etc. sent for chemical bacteriological, histopathological and parasitological examinations.)
A. Blood samples were received from two ailing cases
B. Pieces of spleen and heart

Microscopical examination of blood etc.

1, 2, 3- A few gram positive putrefactive rods.

Results of chemical, bacteriological parasitological and histopathological Examination – A. A marked fall in the leucocyte count i.e. 5000 leucocytes per c. mm of blood)
B. Virological report awaited.

Place where postmortem examination was done ....

Dated-

Sd/- A. Saha

Signature of the autopsist

Designation

Postmortem No. -21

Species – Bovine

Colour – Grey

Owner- A. Prasad

Residence – Patna.

Breed – Red Sindhi

Sex – Female

Age- About 5 –year -old cow

Weight – ....

By whom sent for examination and reasons. If any:

Veterinary Surgeon, Phulwari, Patna.

To ascertain the cause of death.

Date and hour of death/sacrifice	Autopsy findings
10.6.91 at 8.00 a.m	1. Frothy exudate at the mouth.
Date and hour of P.M. exam.	
10.6. 91 at 1.00 p.m	2. Congestion and oedema in the lungs
	3. Subendocardial hemorrhages
	4. Blood tinged contents in the abomasum and intestine.

History :

The symptoms were as under:

Anorexia, excessive salivation, respiratory distress, dilated pupil and muscular twitching following grazing on a paddy field sprayed with dimethoate.

Cadaverous changes- Frothy exudates at the oral orifice.

Rigor mortis –Present in the limbs.

External appearance on removal of skin (describe contusions and wounds etc.)

Mouth and pharynx- Frothy exudate in the mouth and pharynx.

Nasal cavities- Contained frothy exudate

Larynx and trachea –Contained frothy material

Oesophagus –N.A. D.

Pleural cavity and lungs- Haemorrhages on the pleural serosae and subepicardial and subendocardial haemorrhages.

Pericardium and heart (absolute wt..................relative wt....................%)-

Gall bladder

Spleen – N.A. D. (absolute wt...............................relative wt................%)-

Stomach and intestine- Abomasal and intestinal contents tinged with blood and petechiae in the abomasal mucosae.

Urinary organs

Kidneys- The kidneys were congested (absolute wt...............relative wt..............%)-

Generative organs

Brain and spinal cord –Not opened

Lymph glands in general – N. A. D.

Blood or tissue smears sent for-	1. Smears from the heart.
Microscopical examination	2. Smears from the lungs.

Diagnosis on the basis of above examination

Poisoning due to an organ Ophosphorous compound.

Remark (state if viscera and tissue Specimens etc. sent for chemical Bacteriological, histopathological And parasitological examinations) laboratory	A. Pieces of organs and intestinal contents were collected for chemical examination in a forensic

Microscopical examination of blood etc.

1, 2. A few Gram positive rods.

Results of chemical, bacteriological, parasitological and histopathological examinations.

Place where postmortem examination was done....................

Dated –

Note:- Poisoning due to an organopposphorous compound as a presumptive diagnosis was given to be finally confirmed by forensic reports on the content of anti cholesterinases.

Sd/- P. N. Singh

Signature of the autopsist

Designation.

Postmortem No. -22

Species – Canine

Colour – Black

Owner- N. Tiwary

Residence – Patna.

Breed – English bull dog

Sex – Male

Age - About 2-year -old dog

Weight – ....

By whom sent for examination and reasons. If any:

Veterinary Surgeon, Phulwari, Patna.

To ascertain the cause of death.

Date and hour of death/sacrifice	Autopsy findings
12.2.95 at 8.00 a. m.	1. Two injury marks on the muzzle.
Date and hour of P. M. exam.	
12.2.95 at 11.00 a. m.	2. Injured part slightly swollen.

History :

The said dog was bitten by a snake on its nose and showed weak pulse, trembling and paralysis of the limbs before death.

Cadaverous changes – Two marks of haemorrhagic injuries on the nose which was swollen and oedematous . Yellowish conjunctivae of the eyes.

Rigor mortis – Present in the jaw.

External appearance on removal of skin (describe contusions and wounds etc.) N. A. D.

Mouth and pharynx- N. A. D.

Nasal cavities – N.A.D

Larynx and trachea- N. A. D.

Oesophagus – N. A. D.

Pleural cavity and lungs- Pale

Pericardium and heart (absolute wt.....................relative wt........................%)

Partially clotted blood in the ventricles of hearts.

Aortic intima, Pale

Peritoneum

Liver (absolute wt......................relative wt..................%)

Congested liver

Gall bladder,

Spleen - Congested (absolute wt ....relative wt............... %)

Stomach and intestine - Distended with gas.

Urinary organs - N.A. D.

Kidneys (absolute wt..................relative wt.................%)

Congested and slightly red tinged urine in the urinary bladder.

Generative organs N. A. D.

Brain and spinal cord - Opened, Hyperaemic meaninges.

Lymph glands in general

Blood or tissue smears sent for	1. Smears from the heart blood.
Microscopical examination	2. Smears from the liver.

Diagnosis on the basis of above examination.

Snake bite.

Remarks (state if viscera and tissue specimens etc. sent for chemical, bacteriological, histopathological and parasitological examinations).

Microscopical examination of blood etc.

1, 2 - Negative.

Results of chemical, bacteriological , parasitological and histopathological examinations.

Place where postmortem examination was done ....

Dated-

Sd/- P. N. Singh

Signature of the autopsist

Designation

Postmortem No. -23

Species – Equine

Colour – Red

Owner- B. Prasad

Residence – Patna.

Breed – An indigenous mare

Sex – Female

Age- 6 Year - old mare

Weight – ....

By whom sent for examination and reasons. If any:

Veterinary Surgeon, Phulwari, Patna.

To ascertain the cause of death.

Date and hour of death/sacrifice	Autopsy findings
20.5.1992 at 10.00 a.m.	1. Internal hemorrhage following injury to
Dated and hour of P.M. exam. 20.5. 1992 at 3.00 a. m.	muscles, liver and brain by broken bone pieces.
	2. Two litres of blood in the thorax.

History :

A speeding truck hit a horse stading on a road causing fractrure of both hind limbs and injury to the head. The injured animal died after two hours followings the accident.

Cadaverous changes – Present in the jaws. The tibia of both legs were broken in two pieces with spicules of bone protruding through the skin.

External appearance on removal of skin (Describe contusion

and wounds etc.)

Presence of swelling. Around the fractured bones with blood adhering to the skin. The frontal bones of the head were fractured exposing the substance of the brain. The muscles were pale.

Mouth and pharynx- N. A. D.

Nasal cavities - Bloody fluid in the cavities.

Larynx and trachea - Had pale mucosae.

Oesophagus- N.A. D

Pleural cavity and lungs- The first three thoracic ribs on the left side were broken in their middle with penetration of these bones into the left lung. About of 2 litres of blood in the thorax.

Pericardium and heart (absolute wt........................relative wt......................%)

Peritoneum - About one Kg of blood mixed fluid.

Pale. The ventricles of the heart were empty. About 2 litres of blood in the thorax .

Liver (absolute wt.................. relative wt.....................%)

Pale. Last broken thoracic rib on the right side had penetrated into the substance of the liver.

Gall bladder.

Spleen - Pale (absolute wt..............elative wt.................%)

Stomach and intestine- Intestines distended with gas.

Urinary organs

Kidneys - The kidneys were pale (absolute wt...........relative wt............%)

Generative organs- Pale

Brain and spinal cord - About 3 ounces of blood clots in the cerebrum. One inch long bony spicule in the cerebrum

Lymph glands in general - Pale.

Blood or tissue smears sent for – Microscopical examination

1. Smears from the heart
2. Smears from the liver
3. Smears from the lungs

Diagnosis on the basis of above examination –

Secondary shock from internal hemorrhage.

Remarks (state if viscera and tissue specimens etc. sent for chemical, bacteriological, histopathological and parasitological examinations)

Microscopical examination of blood etc.

1, 2 and 3- A few Gram positive rods.

Results of chemical, bacteriological parasitological and histopathological examinations.

Dated –

N. B. : Following a traumatic injury, primary shock occurs in an immediate conjunction to trauma. But the secondary shock occurs 2 to 24 hours of the traumatic injury.

Sd/- P.N. Singh

Signature of the autopsist

Designation

Postmortem No. -24

Species – Caprine

Colour – Black

Owner - Tulsy Roy

Residence – Patna.

Breed – An indigenous stock

Sex – Female

Age - About 6 month -old goat

Weight – ....

By whom sent for examination and reasons. If any:

Veterinary Surgeon, Phulwari, Patna.

To ascertain the cause of death.

Date and hour of death/ sacrifice Autopsy findings

12. 3.91 at 8.00 a. m.	1. Burn injuries on different
Date and hour of P.M. exam	parts of the body .
12.3 .91 at 11.00 a. m.	2. Heat stiffening of the body

History – The goat received burn injuries in a bush fire. Burnt patches of hairs were seen on the thigh and neck.

Cadaverous changes-Patches of hairs and skin over the thigh, neck and back showed carbonaceous material like charcoal.

Rigor mortis - No Rigomortis seen. Presence of stiffened hind limbs.

External appearance on removal of skin (describe contusion and wounds etc.)

Skin and muscles on the jaw bones were charred, dry and destroyed exposing the bones.

Month and pharynx-N.A.D

Nasal cavities- N.A.D

Larynx and trachea

Oesophagus- Haemorrhages on the thoracic pleura

Pleural cavity and lungs- The lungs had dull appearance and moderately congested.

Pericardium and heart (absolute wt.................relative wt.......................%)

Partially clotted blood (about one ounce) in the left ventricle.

Peritoneum – Presence of petechiae on the peritoneal serosae .

Liver – Congested liver(absolute wt..................relative wt..................%)

Gall bladder-N.A. D

Spleen - Slightly enlarged and congested with brownish colour. (absolute wt.....................relative wt.............................%)

Stomach and intestine –Distended with gas. Blood tinged duodenal contents.

Urinary organs Hyperemia and haemorrhages in the adrenal glands.

Kidneys (absolute wt..................relative wt...................%)

Generative organs: N. A. D.

Brain and spinal cord - Meaninges hyperaemic and swollen.

Lymph glands in general – Dull appearance

Blood or tissue smears sent for blood — 1. Smears from the heart

Microscopical examination — 2. Smears from the liver.

Diagnosis on the basis of above examination – **Burn shock.**

Remark (state if viscera and tissue

Specimens etc. sent for chemical,

Bacteriological, histopathological

and parasitological examinations)

Microscopical examination of blood etc.

1, 2- Negative.

Results of chemical, bacteriological

Parasitological and histopathological examinations

Place where postmortem examination was done................

Dated

Noted: In cases of burn injuries, no rigor mortis develops ir thc dead body.

S d/- P. N. Singh

Signature of the autopsist

Designation

Postmortem No. 25

Species- Bovine

Colour- Grey

Owner – A. N. Lal

Breed- Tharparkar

Sex- Male

Age – 6-year old bullock

Weight ....

By whom sent for examination and reasons, if any:

Veterinary Assistant Surgeon.

To find out the cause of death

Date and hour of death / sacrifice	Autopsy findings
3.2.1992 at 9.00 a. m.	1. Singed stumps of hair
Date and hour of P. M. exam.	at the margins of the lacerating wounds in the skin.
3. 2.1992 at 1.00 a.m.	2. Subeutaenous and sub pleural haemorrhages .

History :-

While grazing in the field, the animal was struck by lightning, Restlessness, nervous symptoms and hyperesthesia were seen before its death.

Cadaverous changes - Singed hairs with short stumps on the lateral surface of the thigh.

Rigor mortis- Present in the limbs. Branching lacerations on the left thigh.

External appearance on removal of skin (describe contusions and wounds etc.)

Blood discharge at the nostrils. Several subcutaneous petechiae. Congestive changes in the muscles hit by lightning. Linear hemorrhages in the skin. Presence of poorly clotted dark blood.

Mouth and pharynx- N.A.D

Nasal Cavities- N.A.D

Laryneand trachea- N.A.D

Oesophagus –Do

Pleural cavity and lungs – Haemorrhages on the pleural surface.

Subpleural hemorrhages in the lungs. Oedema of the lungs.

Pericardium and heart (absolute wt................. relative wt........................%)

Partially clotted blood in the left ventricle. Presence of subepicardial and subendocardial haemorrhages.

Peritoneum- Petechiae on the peritoneal serosae.

Liver (absolute wt.................relative wt......................%)

Partly congested and subcapsular haemorrhages

Gall bladder – N.A.D

Spleen –N.A.D (absolute wt ....relative wt................%)

Stomach and intestine- Distended with gas.

Urinary organs-

Kidneys A few subcapsular haemorrhages.

Generative organs-N.A.D

Brain and spinal cord- Opened heamorrhages in the meanings and also on the cut surface of the brain.

Lymph glands in general –N.A.D

Blood or tissue smears sent for- 1.Smears from the lungs.

Microscopical examination.2. Smears from the heart.

Diagnosis on the basis of above examination.

Respiratory arrest following lightning injury.

Remarks (state if viscera and tissue specimens etc. sent for chemical, bacteriological, histopathological and parasitological, examinations.)

Microscopical examination of blood etc.

1, 2- A few Gram positive putrefactive rods.

Results of chemical, bacteriological, parasitological and histopathological examination.

Place where postmortem examination was done ....

Dated

Sd/- P.N. Singh

Signature of the autopsist

Designation

Postmortem No. 26

Species- Bovine

Colour- Grey

Owner – A. Lal

Breed- Haryana

Sex- Female

Age – 5-Year old cow

Weight ....

By whom sent for examination and reasons, if any:

Veterinary Assistant Surgeon, Bihta.

To ascertain the cause of death

Date and hour of death / sacrifice Autopsy findings

12.3.95 at 10.00 a. m. 1. Frothy watery material in the

Date and hour of P. M. exam. trachea and bronchi

12.3.95 at 1.00 a.m. 2. Water in the rumen

History :-

The cow was reported to have fallen into a well and was taken out from the well by the villagers and brought to the laboratory for autopsy.

Cadaverous changes – Abdomen distended with gas. Frothy material at the nostrils.

Rigor mortis – Present in the limbs. Open wound on the left side of the neck. Presence of folds in the skin.

External appearance on removal of skin (describe contusions and wounds etc.) The muscles were pale.

Mouth and pharynx – Frothy material at oral orifice.

Nasal cavities - Frothy material

Larynx and trachea – Contained frothy material.

Oesophagus –N.A. D

Pleural cavity and lungs- The lungs were distended and watery fluid escaped from the cut surfaces.

Pericardium and heart (absolute wt............ relative wt.......................%)

Partially clotted blood in the left ventricle. Dark partially clotted blood in the veins.

Peritoneum – N.A.D

Liver (absolute wt............... relative wt................. %) - partly decomposed.

Gall bladder –N.A. D

Spleen – N.A. D (absolute wt........ relative wt............... %)

Stomach and intestine –T here was a litre of water in the rumen.

An excess of fluid in the abomosum.

Urinary organs

Kidneys (absolute wt.................... relative wt. .... %)

Generative organs – Presence of a 3 month – old foetus in the uterus. Brain and spinal cord – Opened. The cut surface of the brain were pale.

Lymph glands in general – N.A.D

Blood or tissue smears sent for – 1. Smears from the lungs.

Microscopical examination 2. Smears from the heart.

Diagnosis on the basis of above examination

Asphyxia due to drowning in water.

Remark (state if viscera and tissue specimens etc. sent for chemical. Bacteriological, histopathological and parasitological examinations.)

Microscopical Examination of Blood Etc. 1 & 2 – a few gram positive putrefactive rods.

Result of chemical, bacteriological, parasitological and hestopathological examinations.

Place where postmortem examination was done ....

Dated -

Note: Failure to inhale causes, death in 3 to 7 minutes in man but there may be variations in different species.

Sd/- A. Prasad

Signature of the autopsist

Designation.

Postmortem No. 27

Species- Bovine

Colour- Grey

Owner- Cattle Farm, Patna. Breed – Tharparkar

Residence – Patna Sex – Female

Age – about 6-year-old cow

Weight ....

By whom sent for examination and reasons, if any:

Veterinary surgeon.

To ascertain the cause of death.

Date and hour of death/ sacrifice Autopsy findings

10.3.1990 at 2.00 a.m 1. Hypertrophied heart with chronic pericarditis.

Date and hour of P.M. exam. 10.3.1990 at 3.00 a.m 2. Jelly like substance (serous atrophy of fat in the coronary groove of heart.

History: - The said cow was weak and anaemic with swelling in the neck and brisket. It showed frequent tympanitis, anorexia. Prominent neck veins and rise of temperature.

Cadaverous change- Subcutaneous oedema in the neck and sternum. Conjunctivae reddened. Well marked neck veins.

Rigor mortis- Present in the fore limbs.

External appearance on removal of skin (describe contusions and wounds etc.)

The muscles in the sternum infiltrated with jelly like material soaked with water.

Mouth and pharynx – Pale

Nasal cavities – Pale

Larynx and trachea – N.A.D.

Oesophagus - N.A.D.

Pleural cavity and lungs – Congested and consolidated patches in the lungs.

Pericardium and heart – Rounded heart like a football containing 2 Liters of blood tinged fluid with fibrinous flakes. Fibrinous strands were hanging from the thickened pericardium. A nail was found piercing into the pericardium close to the apex in an organized fibrinous tract with a part of the nail into the reticulum. The reticulum was also adhering to the diaphragm.

Peritoneum : About a pound of fluid in the abdominal cavity.

Liver- A capsulated abscess in the liver containing a Il of pus with congested and swollen appearance. (absolute wt......... relative wt ....%)

Gall bladder

Spleen – Congested (absolute wt .... relative wt .... %)

Stomach and intestine – Part of the nail was noticed into reticulum which was attached with the diaphragm by fibrinous adhesions.

Urinary organs – Kidneys congested. The fat between the renal calices was gelatinised.

Generative organsa- N.A. D.

Brain and spinal cord-Not opened

Lymph glands in general-Pale

Blood or tissue smears for 1. Smears from the heart of the cow.

Microscopical examination- 2. Smears from the liver of the cow.

Diagnosis on the basis of above examination.

Traumatic pericardites.

Remarks (state if viscera and tissue specimens etc. sent for chemical, bacteriological, histopathological and parasitlogical examination)

Microscopical examination of blood etc.

1 and 2- Gram positive rods.

Results of chemical , bacteriological

Parasitological and histopathological

Examinations

Place where postmortem examination was done.......................
Dated

Sd/- A. Prasad

Signature of the autopsist

Designation-

Postmortem No- 28..............................

Species Bovine

Colour Reds Breed Red Sindhi

Owner Neha Sex Female

ResidenceAshok Nagar, k-bagh Patna-20 Age About 1-year-old

Weight............

By whom sent for examination and reasons if any Veterinary Surgeon

Date and hour of death / sacrifice 12-01-2008 at 3 a.m

Autopsy findings

1. Anaemia
2. Abomasitis and enteritis
3. Gelatinised tissues in the marrow cavities and coronary groove of heart.
4. Several trichostongylids in the abomasum.

Date and hour of postmortem examination 12-01-2008 at 10. A.M

History - As reported by the owner, the calf was debilitated, suffered from cachexia, diarrhea and developed swelling under the jaw bone. Dark colored foetid matter was defecated by the calf.

Cadaverous changes - Poor nutritional condition with emaciation and pale conjuncetivae. Oedmatous swelling in the subtaneous tissue under the jaw and abdomen subcutaneously.

Rigor mortis Present in only hind limbs

External appearance on removal of

skin (describe contusions and wounds etc.) The muscles are pale. The blood was thin and watery and did not stain the fingers tips

Mouth and pharynx Pale mucosae

Nasal cavities NAD

Larynx and trachea Pale mucosae

Oesophagus NAD

Pleural cavity and lungs Both diaphragmatic lobes of the lungs showed reddened areas which released blood from the cut surfaces

Pericardium and heart (absolute wt....... relative wt..........%
Partially clotted blood in the left ventricle. There was jelly like material in the coronary groove of heart. The muscle of heart was pale

Peritoneum NAD

Liver (sbsolute wt........relative wt..............................%
The liver was pale and swollen and friable

Gall bladder Thin watery bile in the gallbladder

Spleen (absolute wt..................relative wt..................%
Pale and wrinkled capsule

Stomach and intestine

The mucose of the abomasum and duodenum was reddened and swollen and covered with mucoid slightly blood tinged material. When the abomasal contents were sieved through, large number of round worms where detected. Such worms were also present in the duodenal contents.

Urinary organs

Kidneys (absolute wt ....relative wt..................%
Pale and anaemic

Generative organs NAD

Brain and spinal cord Not opened

Lymph glands in general Pale and anaemic. When the bones were sawn, there was creamed coloured gelatinized material on the cut surfaces of the bone marrow.

Blood or tissue smears sent for microscopical examination

1. Blood smears from the heart

2. Smears from the lungs.

3. Examination of the abomasal contents

Diagnosis on the basis of above examination

Trichostongyliasis

Remarks (state if viscera and tissue specimens etc. sent for chemical, histopathological and parasitological examinations)

Microscopical examination of blood etc.

Results of chemical, bacteriological parasitological and histo pathological examinations 1& 2 negative

3. Trichostrongyles and their eggs in the faeces were detected

Place where postmortem examination was done ....

Dated :

Sd/- B.K. Sinha

Signature of the autopsist

Designation :

Postmortem No- 29............................

Species Canine

Colour Breed Alsatian

Owner Nibha Sex Female

Residence.........................Age About 2- month- old Weight............

By whom sent for examination and reasons if any

Veterinory surgeon To ascertain the cause of the death.

Date and hour of death / sacrifice 29-01-2008 at 6: a.m

Date and hour of postmortem Examination. 29-01-08 at 3 P.M	Autopsy findings 1. Broncho-pneumonia 2. Gestro enteritis 3. Pharyngitis 4. Conjunctivitis

History- As reported by the owner, the said pet showed rise of temperature (105°F) , muco purlulent discharge from the nose and eyes, salivation, diarrhoea epileptic seizure and quiverings in the limbs. A biphasic fever (i.e. the first rise of body temperature followed by a second rise of temperature) was reported.

Cadaverous changes- The caracase was emaciated and showed with poor state of nutrition. Conjunctivae of the eyes were reddened with muco purulent discharge.

Rigor mortis - Present in the fore limbs

External appearance on removal of skin (describe contusions and wounds etc.)- The muscles were dry, sticky to touch and reduced in bulk.

Mouth and pharynx- Purulent mucoid cream coloured material over the pharyngeal mucose.

Nasal cavities- Mucoid thick material over nasal mucosae

Larynx and trachea- The tracheal mucosa was reddened.

Oesophagus- NAD

Pleural cavity and lungs- Both the lungs were reddened and consolidated in patches. The cut surfaces the of the lungs revealed the bulging appearance. Firm or solid pieces of the lungs sank into water in a pot.

Pericardium and heart (absolute wt....... relative wt..........%- Some clotted blood in the ventricles of the heart

Peritoneum- NAD

Liver (sbsolute wt........relative wt.............................. %- Swollen and moderately reddened

Gall bladder- Slightly thick bile in the gall bladder

Spleen (absolute wt.................relative wt..................%- Reddened and enlarged in size

Stomach and intestine – The mucoase of the stomach and duodenum were reddened and covered with slightly blood tinged material.

Urinary organs

Kidneys (absolute wt ....relative wt..................%- Enlarged and moderately reddened. The mucoase of the urinary bladder was slightly reddened.

Generative organs-NAD

Brain and spinal cord- Not opened

Lymph glands in general

Blood or tissue smears sent for microscopical examination-

1. Smears from the heart
2. Smears of the lungs.

Diagnosis on the basis of above examination- Canine distemper

Remarks (state if viscera and tissue specimens etc. sent for-

1. Pieces of tissues from the nasal cavities and mucosae of the urinary bladder were preserved in 10 percent formalin solution.

chemical, histopathological and parasitological examinations)

Microscopical examinationof blood etc.

Results of chemical, bacteriological parasitological and histopathological examinations- 1 and 2 A few Gram negative organisms

3. Intrascytoplasmic and intranucleor inclusions in the scrapings or sections of the nasal or bladder mucosal tissues stained by Schorr- 3.

Place where postmortem examination was done ....

Dated :

Signature of the autopsist

Designation :

Postmortem No-30.............................

Species - Bovine

Colour.....................................Breed - Tharparker

Owner Archana Sex- Female

Residence.................................Age-3- year- old

Weight............

By whom sent for examination and reasons if any-

Veterinary Doctor

To as certain the cause of death.

Date and hour of death / sacrifice 28-01-08 at 2:a.m

Autopsy findings

i. Dermatitis with slough of skin in patches.

ii. Congested and consolidated patches

iii. Pale and swollen kidneys

Date and hour of postmortem examination 28.01.08 at 9. A.M

History - As per history of the case, the animal was slated to have eaten lantana weeds in the hedges of the compounds in the Kankarbagh area. Peels and slough of the skin with escape of fluid were noticed during life. The patient preferred to stand under tree shade to avoid the sun light.

Cadaverous changes- The skin sowed swelling, loss of superficial layers in skin in the unpigmented area s of the body.
Rigor mortis- Present in the limbs marked by difficulty in flexing the joints of limbs.

External appearance on removal of

skin (describe contusions and wounds etc.) Muscles pale and dry and the conjunctivae of the eyes were yellow. Presence of peels of the skin, loss of hairs reddened or grayish and swollen exposed dermis

Mouth and pharynx-Yellow

Nasal cavities-Pale

Larynx and trachea-Pale

Oesophagus-NAD

Pleural cavity and lungs-Diaphragmatic lobes of the lungs were reddened and consolidated in patches. The cut pieces of the lungs showed dry raised areas and such farm pieces sank into water.

Pericardium and heart (absolute wt....... relative wt..........%- Some clotted blood in the left ventricle of the heart.

Peritoneum- NAD

Liver (sbsolute wt........relative wt................................%- The liver was swollen and its cut surfaces revealed yellowish brown discolouration.

Gall bladder-Thick ropy bile

Spleen (absolute wt.................relative wt................/%-Pale

Stomach and intestine –Hemorrahagic mucoid material on the gut mucosae

Urinary organs-NAD

Kidneys (absolute wt ....relative wt..................%-Pale and swollen kidneys.

Generative organs-NAD

Brain and spinal cord- Not opened

Lymph glands in general- Pale

Blood or tissue smears sent for microscopical examination-1.Smears from the liver.

2. Smears from the heart.

Diagnosis on the basis of above examination –Photo sensitization cum bronchopneumonia.

Remarks (state if viscera and tissue specimens etc. sent for chemical, histopathological and parasitological examinations)

Microscopical examinationof blood etc.

Results of chemical, bacteriological parasitological and histo-pathologicalexaminations- 1 & 2 Negative

Place where postmortem examination was done ....

Dated :

Sd/- Alka Saran

Signature of the autopsist

Designation :

References

1. Alpern, D. (1967) Pathologic Physiology, Mir Publishers, Moscow.
2. Cotran Ram ji S; Kumar, Robbins: Pathologic Vinay and Collins Tucker (1999) basis of diseases 6th edit, W.B. Saunders Company, Harcourt Publishers, International Company
3. Davies, G.O. (1960) Veterinary Pathology and Bacteriology Lioyd- Luke (Medical Books) Ltd., 49 Newman Street, London
4. Enzinger, Franz, M and Soft tissue tumours, M, mosby Sharou, Weiss, W. (1995) St. Louis, Missauri.
5. Gupta, U.K. and Prasad, B.N. (Personal communication) Medical officer Health services, Bihar.
6. Herbut, Pater A. (1959) Pathology Lea and Febiger, Philadelphia.
7. Jones, T.C. and Hunt, R.D. (1983) Veterinary Pathology, Lea and Febiger, Philadelphia.
8. Kalra, D.S. (1971) How to collect and prepare materials for dispatch to the laboratory. Department of Pathology, Punjab Agricultural University, Hissar.
9. Prasad, Mohan and Singh, Vinod Kumar.Personal communication. Head and Controlling Officer/ Principal, Bihar college of Physiotherapy and Occupational therapy, P.M.C.H. campus, Patna and Head Department of Nephrology, P.M.C.H. Patna
10. Rubarth, S, and Hansen Some aspects of General Pathology. H.J. (1964) 2nd F.A.O./Swedish Training Centre in Post Graduate Veterinary Pathology, Stockholm Sweden.
11. Runnel et. al. (1960) Principles of Veterinary Pathology, the Iowa State University Press, Ames, Iowa, U.S.A.
12. Sharma, S.N. (1981) Veterinary Jurisprudence. Oxford & I.B.H. Publishing Co., New Delhi.
13. Singh, Shanti, Singh, S.B., and Prasad, L.N. and Sinha, S.R.P. Former Associate Professor (Gyanecology), P.M.C.H. Patna, Former Director, Health Services Government of Bihar and Head

Department of Pathology, B.V.C. Patna. (Personal Communication).

14. Stevens, A.J. (1961) A laboratory hand book of Veterinary diagnostic methods Animal Health branch monogram. No. 3, FAO, Rome.
15. Willis, R.A. (1960) Pathology of Tumours, 3rd Edit. London, Butterworth & Co.

Index

Fig. 1 and 2. Sitting (third from left), the former Cabinet Minister of Social Justice and Empowerment but Presently the Lok Sabha Speaker of Indian Parliament, Government of India during the release of ' Advanced Pathology and Treatment of Diseases of Domstic Animals' written by Dr. C.D.N.Singh. ex- University Professor of Pathology, Bihar Veterinary College Patna,Rajendra Agriculture University on 8th. July,2008